OFFICIAL
PAST
PAPERS
WITH ANSWERS

HIGHER

BIOLOGY
2006-2009

© Scottish Qualifications Authority
All rights reserved. Copying prohibited. No part of this publication may be reproduced, stored in a retrieval system, or transmitted in any form or by any means, electronic, mechanical, photocopying, recording or otherwise.

First exam published in 2006.
Published by Bright Red Publishing Ltd, 6 Stafford Street, Edinburgh EH3 7AU
tel: 0131 220 5804 fax: 0131 220 6710 info@brightredpublishing.co.uk www.brightredpublishing.co.uk

ISBN 978-1-84948-053-6

A CIP Catalogue record for this book is available from the British Library.

Bright Red Publishing is grateful to the copyright holders, as credited on the final page of the book, for permission to use their material. Every effort has been made to trace the copyright holders and to obtain their permission for the use of copyright material. Bright Red Publishing will be happy to receive information allowing us to rectify any error or omission in future editions.

HIGHER

2006

[BLANK PAGE]

FOR OFFICIAL USE

Total for
Sections
B and C

X007/301

NATIONAL
QUALIFICATIONS
2006

TUESDAY, 23 MAY
1.00 PM – 3.30 PM

BIOLOGY
HIGHER

Fill in these boxes and read what is printed below.

Full name of centre

Town

Forename(s)

Surname

Date of birth
   Day   Month   Year   Scottish candidate number   Number of seat

SECTION A—Questions 1–30 (30 marks)

Instructions for completion of Section A are given on page two.

For this section of the examination you must use an **HB pencil**.

SECTIONS B AND C (100 marks)

1   (a)   All questions should be attempted.

      (b)   It should be noted that in **Section C** questions 1 and 2 each contain a choice.

2   The questions may be answered in any order but all answers are to be written in the spaces provided in this answer book, **and must be written clearly and legibly in ink**.

3   Additional space for answers will be found at the end of the book. If further space is required, supplementary sheets may be obtained from the invigilator and should be inserted inside the **front cover** of this book.

4   The numbers of questions must be clearly inserted with any answers written in the additional space.

5   Rough work, if any should be necessary, should be written in this book and then scored through when the fair copy has been written. If further space is required a supplementary sheet for rough work may be obtained from the invigilator.

6   Before leaving the examination room you must give this book to the invigilator. If you do not, you may lose all the marks for this paper.

SCOTTISH
QUALIFICATIONS
AUTHORITY

Read carefully

1   Check that the answer sheet provided is for **Biology Higher (Section A)**.

2   For this section of the examination you must use an **HB pencil**, and where necessary, an eraser.

3   Check that the answer sheet you have been given has **your name**, **date of birth**, **SCN** (Scottish Candidate Number) and **Centre Name** printed on it.

Do not change any of these details.

4   If any of this information is wrong, tell the Invigilator immediately.

5   If this information is correct, **print** your name and seat number in the boxes provided.

6   The answer to each question is **either** A, B, C or D. Decide what your answer is, then, using your pencil, put a horizontal line in the space provided (see sample question below).

7   There is **only one correct** answer to each question.

8   Any rough working should be done on the question paper or the rough working sheet, **not** on your answer sheet.

9   At the end of the exam, put the **answer sheet for Section A inside the front cover of this answer book**.

Sample Question

The apparatus used to determine the energy stored in a foodstuff is a

A   calorimeter

B   respirometer

C   klinostat

D   gas burette.

The correct answer is **A**—calorimeter. The answer **A** has been clearly marked in **pencil** with a horizontal line (see below).

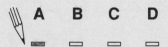

Changing an answer

If you decide to change your answer, carefully erase your first answer and using your pencil fill in the answer you want. The answer below has been changed to **D**.

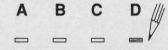

SECTION A

All questions in this section should be attempted.

Answers should be given on the separate answer sheet provided.

1. The diagram below represents a highly magnified section of a yeast cell.

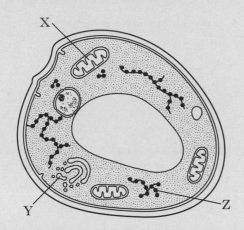

Which line of the table below correctly links each cell structure with its function?

	Aerobic respiration	Protein synthesis	Packaging materials for secretion
A	X	Y	Z
B	Y	Z	X
C	X	Z	Y
D	Z	X	Y

2. The action spectrum in photosynthesis is a measure of the ability of photosynthetic pigments to

   A absorb red and blue light

   B absorb light of different intensities

   C carry out photolysis

   D use light of different wavelengths for synthesis.

3. The diagram below shows the energy flow in an area of forest canopy during 1 year.

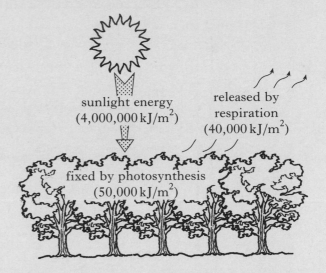

What percentage of available sunlight energy is fixed by the trees?

   A 0·25%

   B 1·00%

   C 1·25%

   D 2·25%

[Turn over

4. Photosynthetic pigments can be separated by means of chromatography as shown in the diagram below.

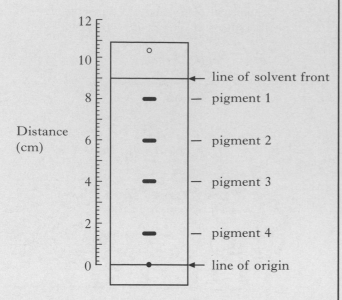

A pigment can be identified from its R_f value which can be calculated as follows:

$$R_f = \frac{\text{distance travelled by pigment from origin}}{\text{distance travelled by solvent from origin}}$$

Which line of the table correctly identifies the R_f values of pigments 2 and 3 on the above chromatogram?

	Pigment 2	*Pigment 3*
A	0·44	0·17
B	0·67	0·44
C	0·44	0·67
D	0·67	0·89

5. The diagram below shows a mitochondrion surrounded by cytoplasm.

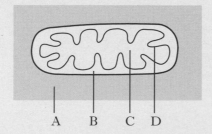

Where does glycolysis take place?

6. Which of the following statements refer to glycolysis?

 1   Carbon dioxide released.

 2   Occurs during aerobic respiration.

 3   The end product is pyruvic acid.

 4   The end product is lactic acid.

 A   1 and 3

 B   1 and 4

 C   2 and 3

 D   2 and 4

7. During anaerobic respiration in muscle fibres what is the fate of pyruvic acid?

 A   It is converted to lactic acid.

 B   It is broken down by the mitochondria.

 C   It is broken down to carbon dioxide and water.

 D   It is converted to citric acid.

8. Which of the following proteins has a fibrous structure?

 A   Insulin

 B   Pepsin

 C   Amylase

 D   Collagen

9. If ten percent of the bases in a molecule of DNA are adenine, what is the ratio of adenine to guanine in the same molecule?

 A   1:1

 B   1:2

 C   1:3

 D   1:4

10. In the life cycle of a bacterial virus which of the following sequences of events occurs?

 A   Lysis of the cell membrane, synthesis of viral DNA, replication of viral protein

 B   Lysis of cell membrane, synthesis of viral protein, replication of viral DNA

 C   Replication of viral DNA, synthesis of viral protein, lysis of cell membrane

 D   Synthesis of viral protein, replication of viral DNA, lysis of cell membrane

11. The table below shows some genotypes and phenotypes associated with forms of sickle-cell anaemia.

Phenotype	Genotype
unaffected	$Hb^A Hb^A$
sickle-cell trait	$Hb^A Hb^S$
acute sickle-cell anaemia	$Hb^S Hb^S$

A woman with sickle-cell trait and a man who is unaffected plan to have a child.

What are the chances that their child will have acute sickle-cell anaemia?

A None

B 1 in 1

C 1 in 2

D 1 in 4

12. The following cross was carried out using two pure-breeding strains of the fruit fly, *Drosophila*.

P   straight wing   curly wing
              +   ×   +
           black body   grey body

F_1 All straight wing + black body

    The F_1 were allowed to interbreed

F_2 straight wing   curly wing
              +   ×   +
           black body   grey body

F_2 Phenotype
Ratio   3   :   1

In a dihybrid cross the typical F_2 ratio is $9:3:3:1$.

An explanation of the result obtained in the above cross is that

A crossing over has occurred between the genes

B before isolation F_1 females had mated with their own type males

C non-disjunction of chromosomes in the sex cells has taken place

D these genes are linked.

13. The recombination frequency obtained in a genetic cross may be used as a source of information concerning the

A genotypes of the recombinant offspring

B diploid number of the species

C fertility of the species

D position of gene loci.

14. Red-green colour deficient vision is a sex-linked condition. John, who is affected, has the family tree shown below.

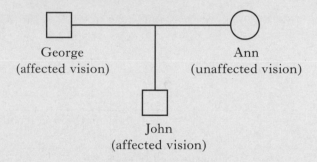

George
(affected vision)

Ann
(unaffected vision)

John
(affected vision)

If b is the mutant allele for the condition, which of the following could be the genotypes of George's parents and Ann's parents?

	George's parents		Ann's parents	
A	$X^B X^b$	$X^B Y$	$X^B X^B$	$X^B Y$
B	$X^B X^B$	$X^b Y$	$X^B X^B$	$X^B Y$
C	$X^B X^b$	$X^B Y$	$X^B X^b$	$X^B Y$
D	$X^B X^B$	$X^b Y$	$X^B X^B$	$X^b Y$

15. Klinefelter's syndrome is caused by the presence of an extra X chromosome in human males. Affected individuals are therefore XXY.

This syndrome is caused by

A recombination

B sex-linkage

C crossing-over

D non-disjunction.

[Turn over

16. The table refers to the mass of DNA in certain human body cells.

Cell type	Mass of DNA in cell ($\times 10^{-12}$ g)
liver	6·6
lung	6·6
R	3·3
S	0·0

Which of the following is the most likely identification of cell types R and S?

	R	S
A	ovum	mature red blood cell
B	mature red blood cell	sperm
C	nerve cell	mature red blood cell
D	kidney tubule cell	ovum

17. The following steps are involved in the process of genetic engineering.

1 Insertion of a plasmid into a bacterial host cell

2 Use of an enzyme to cut out a piece of chromosome containing a desired gene

3 Insertion of the desired gene into the bacterial plasmid

4 Use of an enzyme to open a bacterial plasmid

What is the correct sequence of these steps?

A   4   1   2   3

B   2   4   3   1

C   4   3   1   2

D   2   3   4   1

18. Osmoregulation in bony fish is achieved by a variety of strategies, depending on the nature of the environment.

Strategies

1 Large volume of dilute urine produced.

2 Small volume of concentrated urine produced.

3 Kidneys contain many large glomeruli.

4 Kidneys contain few small glomeruli.

Which of these strategies are employed by a freshwater bony fish?

A   1 and 4

B   2 and 4

C   1 and 3

D   2 and 3

19. The list below shows benefits which an animal species can obtain from certain types of social behaviour.

1 Aggression between individuals is controlled.

2 Subordinate animals are more likely to gain an adequate food supply.

3 Experienced leadership is guaranteed.

4 Energy used by individuals to obtain food is reduced.

Which statements refer to co-operative hunting?

A   1 and 2 only

B   1 and 3 only

C   2 and 4 only

D   3 and 4 only

20. The rates of carbon dioxide exchange by the leaves of two species of plants were measured at different light intensities.

The results are shown in the graph below.

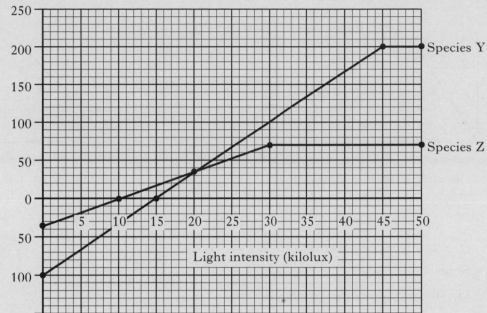

By how many kilolux is the compensation point for species Y greater than the compensation point for species Z?

A   5

B   10

C   15

D   130

21. A feature of phenylketonuria in humans is

A   the synthesis of excess phenylalanine

B   an inability to synthesise phenylalanine

C   the synthesis of excess tyrosine from phenylalanine

D   an inability to synthesise tyrosine from phenylalanine.

22. Plant ovary wall cells develop differently from plant phloem cells because of

A   random assortment in meiosis

B   genes being switched on and off during development

C   their having different numbers of chromosomes

D   their having different sets of genes.

[Turn over

23. An investigation was carried out into the germination of barley seeds.

The concentration of amylase and the rate of breakdown of starch was measured over 15 days.

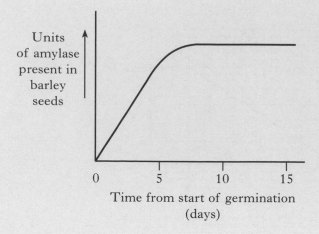

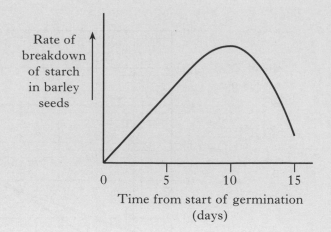

From the graphs it can be seen that after 10 days

A   the production of gibberellin has ceased

B   the rate of starch digestion decreases

C   the barley is synthesising its own starch

D   the amylase is becoming denatured.

---

24. The graph below contains information about fertiliser usage.

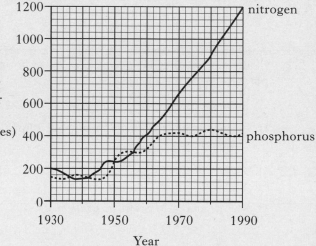

Which of the following statements about nitrogen usage between 1930 and 1990 is correct?

A   It increased steadily.

B   It increased by 500%.

C   It increased by 600%.

D   It always exceeded phosphorus usage.

25. Dietary deficiency of vitamin D causes rickets.

This effect is due to

A   poor uptake of phosphate into growing bones

B   low vitamin D content in the bones

C   poor calcium absorption from the intestine

D   loss of calcium from the bones.

26. Which of the following would result from an increased production of ADH?

A   The production of urine with a higher concentration of urea

B   Decrease in the permeability of the kidney collecting ducts

C   A decrease in the rate of glomerular filtration

D   An increase in the rate of production of urine

27. The graph below records the body temperature of a woman during an investigation in which her arm was immersed in water.

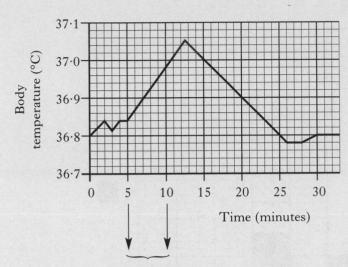

Arm immersed
in water during
this period

By how much did the temperature of her body vary during the 30 minutes of the investigation?

A   2·7 °C

B   0·27 °C

C   2·5 °C

D   0·25 °C

28. The graph below contains information about the birth rate and death rate in Mexico.

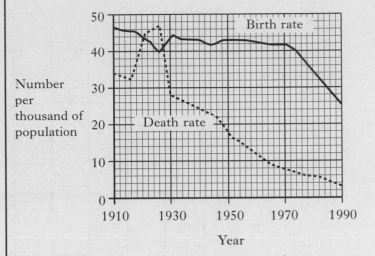

Which of the following conclusions can be drawn from the graph?

A   At no time during the century has the population of Mexico decreased.

B   The greatest increase in population occurred in 1970.

C   The population was growing faster in 1910 than in 1990.

D   Birth rate decreased between 1970 and 1990 due to the use of contraception.

[Turn over

29. The bar chart below shows the percentage loss in yield of four organically grown crops as a result of the effects of weeds, disease and insects.

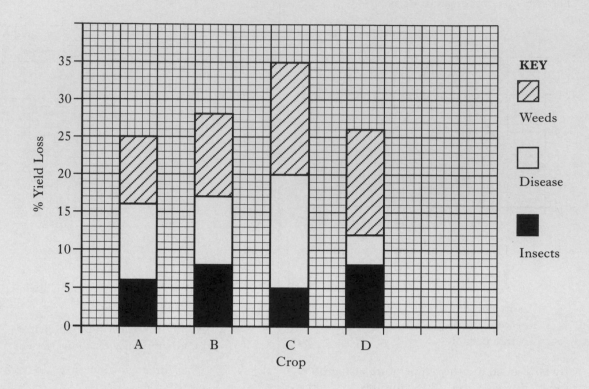

Predict which crop is most likely to show the greatest increase in yield if herbicides and insecticides were applied.

30. During succession in plant communities a number of changes take place in the ecosystem. Which line of the table correctly describes some of these changes?

	Species diversity	Biomass	Food web complexity
A	rises	rises	rises
B	rises	falls	rises
C	falls	rises	rises
D	rises	rises	falls

Candidates are reminded that the answer sheet MUST be returned INSIDE the front cover of this answer book.

[Turn over for Section B on *Page twelve*

Marks

SECTION B

All questions in this section should be attempted.

All answers must be written clearly and legibly in ink.

1. (a) The grid contains information about the plasma membrane and the cell wall.

A contains phospholipid	B fully permeable	C made of fibres
D contains cellulose	E selectively permeable	F made up of two layers

(i) Two of the boxes contain information about the **structure** of the plasma membrane.

Identify these **two** boxes.

Letters _____ and _____   1

(ii) One of the boxes contains information that relates to the role of the cell wall in the movement of water into a cell.

Identify this box.

Letter _____   1

(b) The table shows the percentage change in mass of apple tissue after immersion in sucrose solutions of different concentrations.

Concentration of sucrose solution (M)	Change in mass of apple tissue (%)
0·00	+22·0
0·10	+13·0
0·15	+8·5
0·20	+4·0
0·25	−0·5
0·30	−5·0
0·40	−14·0
0·50	−23·0

Marks

1. **(*b*)** **(continued)**

(i) Using values from the table, plot a line graph to show the percentage change in mass of the apple tissue against the concentration of sucrose solution.

Use appropriate scales to fill most of the grid.

(An additional grid, if required, may be found on page 40.)

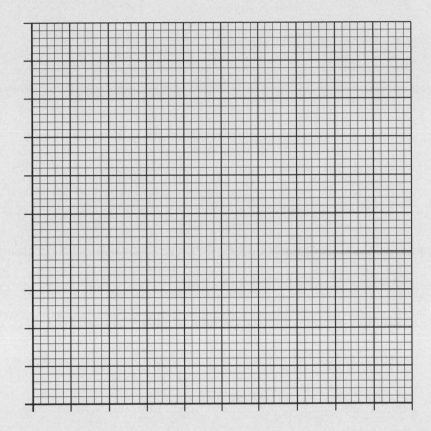

2

(ii) Complete the following sentences by underlining **one** of the alternatives in each pair.

The 0·1 M sucrose solution is $\left\{ \begin{array}{l} \text{hypotonic} \\ \text{hypertonic} \end{array} \right\}$ to the apple tissue.

Apple cells immersed in this solution may become $\left\{ \begin{array}{l} \text{plasmolysed} \\ \text{turgid} \end{array} \right\}$.

1

[Turn over

Marks

2. The diagram shows an outline of the stages in photosynthesis.

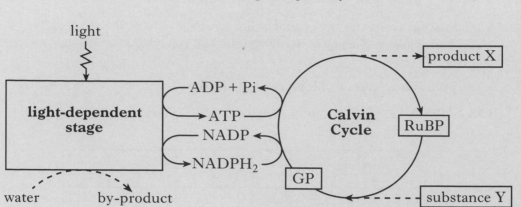

(a)   (i)   Name the by-product produced by the light-dependent stage.

_____ 1

(ii)   Name product X and substance Y.

Product X _____

Substance Y _____ 1

(iii)   State the number of carbon atoms in RuBP and GP.

RuBP _____ GP _____ 1

(b)   (i)   Describe the role of ATP in photosynthesis.

_____

_____

_____ 1

(ii)   Explain why hydrogen from the light-dependent stage of photosynthesis is needed by the Calvin cycle.

_____

_____

_____ 1

Marks

2. **(continued)**

(c) The graph shows the absorption spectra of three photosynthetic pigments.

Pigments P and Q were extracted from a hydrophyte with leaves that float on the water surface.

Pigment R was extracted from a species of photosynthetic algae that lives in the water below the hydrophyte.

——— pigment P
- - - - - pigment Q
·············· pigment R

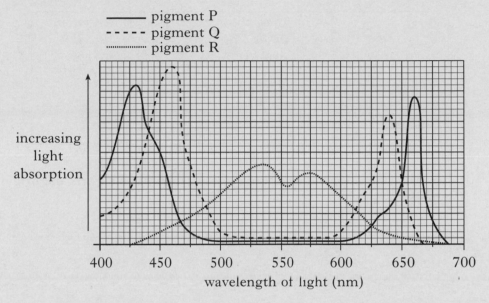

(i) Tick (✓) **one** box to identify the wavelengths of light at which pigment P shows greatest absorption.

425–450 nm	450–475 nm	525–550 nm	625–650 nm	650–675 nm

1

(ii) Explain why it is an advantage to the hydrophyte to have more than one pigment.

_____

_____

_____

1

(iii) Give **one** adaptation of hydrophyte leaves and state its effect.

Adaptation _____

Effect _____

_____

1

(iv) From the information given, explain how the algae from which pigment R was extracted are adapted to photosynthesise in their environment.

_____

_____

_____

1

Marks

3. Experiments were carried out to investigate the hypothesis that the uptake of ions into mammalian cells takes place by active transport.

(a) The concentrations of potassium ions and chloride ions inside and outside a mammalian cell were measured.

The table shows the results obtained at an oxygen concentration of 4·0 units.

Ion	Ion concentration inside cell (mM)	Ion concentration outside cell (mM)
Potassium	140	5
Chloride	10	110

(i) Describe the information shown in the table that supports the original hypothesis.

_____

_____

_____ 1

(ii) From the table, calculate the simplest whole number ratio of potassium ions to chloride ions outside a mammalian cell.

Space for calculation

_____ potassium ions : _____ chloride ions 1

Marks

3. (continued)

(*b*) The graph shows the effect of changing oxygen concentration on the concentration of potassium ions inside a mammalian cell.

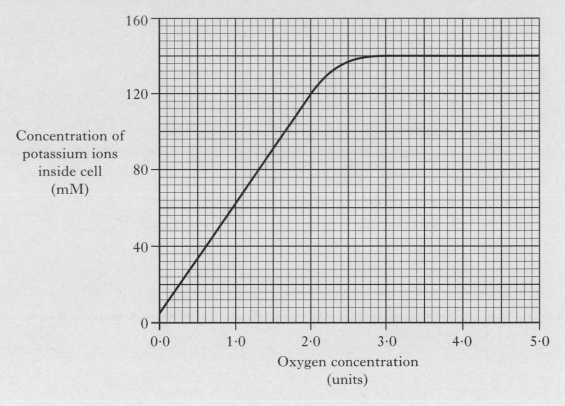

(i) Explain the shape of the graph between oxygen concentrations of 1·0 and 2·0 units.

_____

_____

_____

_____

2

(ii) Suggest a reason why the graph levels off at oxygen concentrations above 3·0 units.

_____

_____

1

[Turn over

Marks

4. Yeast is a micro-organism capable of both aerobic and anaerobic respiration.

(*a*) Describe the role of oxygen in aerobic respiration.

_____

_____   1

(*b*) The diagrams represent a mitochondrion from a normal yeast cell and one from a mutant yeast cell.

mitochondrion from
normal yeast cell

mitochondrion from
mutant yeast cell

(i) Name the structures that are absent from the mitochondrion of the mutant yeast cell.

_____   1

(ii) An experiment was carried out to investigate the effect of oxygen on the growth of normal and mutant yeast cells.

The method used in the experiment is outlined below.

- Three normal yeast cells were placed on agar growth medium containing glucose in a petri dish.

- Three mutant yeast cells were placed on the same agar growth medium containing glucose in a second dish.

- The dishes were then incubated at 30 °C for four days in aerobic conditions to allow the cells to multiply and produce colonies of yeast cells.

- The above three steps were repeated and the dishes were incubated this time in anaerobic conditions.

The sizes of the colonies produced are shown in the following diagrams.

Marks

4. **(b)** **(ii)** **(continued)**

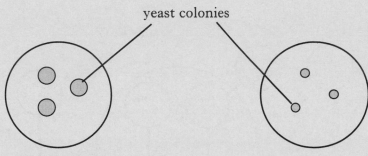

normal yeast
grown in aerobic conditions

mutant yeast
grown in aerobic conditions

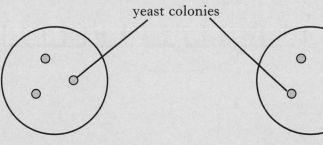

normal yeast
grown in anaerobic conditions

mutant yeast
grown in anaerobic conditions

1. Suggest a possible improvement to the experimental method, other than repeating the experiment, that would increase the reliability of the results.

_____

_____ 1

2. Give an explanation for the difference in colony size observed for the normal yeast grown in aerobic and anaerobic conditions.

_____

_____

_____ 2

3. Suggest why there was no difference in colony size when the mutant cells were grown in aerobic and anaerobic conditions.

_____

_____ 1

[Turn over

Marks

5. The diagram shows translation of part of a mRNA molecule during the synthesis of a protein.

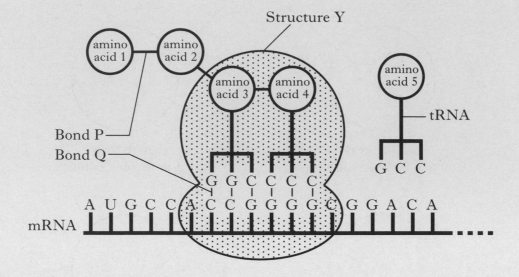

(a) Name structure Y.

_____   1

(b) Name the types of bond shown at P and Q.

Bond P _____

Bond Q _____   1

(c) Give the anticodon for the tRNA for amino acid 1.

_____   1

(d) Describe **two** functions of tRNA in protein synthesis.

1 _____

_____   1

2 _____

_____   1

(e) Genetic information for protein synthesis is in the form of a triplet code. Explain what is meant by this statement.

_____

_____

_____   1

Marks

6. Patients who have had tissue transplants may be treated with a drug that suppresses the immune system.

The table shows the number of lymphocytes in the blood of a patient before and after treatment.

Number of lymphocytes before treatment (cells per mm^3)	Number of lymphocytes after treatment (cells per mm^3)
7500	3000

(a) Calculate the percentage decrease in the number of lymphocytes following treatment with the suppressor drug.

Space for calculation

_____ %   **1**

(b) Explain why there is a risk of rejection when tissues are transplanted.

_____

_____

_____

_____   **2**

(c) Some suppressor drugs act by binding to DNA molecules in such a way that the separation of the two DNA strands is prevented.

Predict **one** possible consequence of the use of this type of suppressor drug on the normal functions of DNA.

_____

_____

_____   **1**

[Turn over

Marks

7. (*a*) The letters A – E represent five statements about meiosis.

Letter	Statement
A	haploid gametes are produced
B	chiasmata are formed
C	chromatids separate
D	gamete mother cell is present
E	homologous chromosomes form pairs

(i) Use **all** the letters from the list to complete the table to show the statements that are connected with the first and the second meiotic divisions.

First meiotic division	*Second meiotic division*

2

(ii) Independent assortment of chromosomes during meiosis is a source of genetic variation.

Describe the behaviour of chromosomes during the first meiotic division stage that results in independent assortment.

_____

_____

_____

1

Marks

7.   (continued)

(b)   Duchenne muscular dystrophy (DMD) is a recessive, sex-linked condition in humans that affects muscle function.

The diagram shows an X-chromosome from an unaffected individual and one from an individual with DMD.

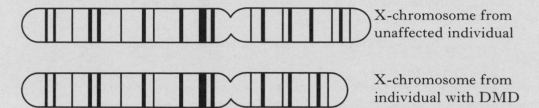

X-chromosome from unaffected individual

X-chromosome from individual with DMD

(i)   Using information from the diagram, name the type of chromosome mutation responsible for DMD.

_____   1

(ii)   **On the diagram** of the chromosome from the **unaffected individual**, put a **cross (X)** on the likely location of the gene involved in DMD.   1

(iii)   Males are more likely to be affected by DMD than females.

Explain why.

_____

_____

_____

_____   1

(iv)   A person with DMD has an altered phenotype compared with an unaffected individual.

Explain how an inherited chromosome mutation such as DMD may result in an altered phenotype.

_____

_____

_____

_____   1

[Turn over

Marks

8. Three species of small bird, the Blue Tit, the Great Tit and the Marsh Tit, forage for caterpillars in oak woods in early summer.

Investigators observed each bird species for ten hours. They recorded the percentage of time each species spent foraging in different parts of the trees. The results are shown in the **Bar Chart**.

Table 1 shows the percentage of the birds' diet that came from different caterpillar size ranges.

Table 2 shows the beak size index calculated for each species using the following formula.

beak size index = average beak length (mm) × average beak depth (mm)

Bar Chart

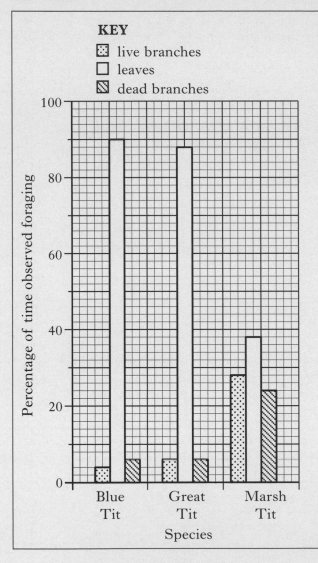

KEY
- ▨ live branches
- ☐ leaves
- ▨ dead branches

Table 1

caterpillar size range (mm)	% of diet from each size range		
	Blue Tit	Great Tit	Marsh Tit
1–2	63·6	18·2	18·1
3–4	24·3	20·4	60·6
5–6	10·0	27·2	12·1
7–8	2·1	34·2	9·2

Table 2

species	beak size index (mm²)
Blue Tit	40·3
Great Tit	67·6
Marsh Tit	44·2

(a) Calculate the number of minutes the Blue Tits were observed foraging on live branches.

Space for calculation

_____ mins   1

8. **(continued)**

Marks

(b) What evidence from the **Bar Chart** suggests that Marsh Tits forage on tree parts other than those shown?

_____

_____   1

(c) Calculate the average percentage of the birds' diets in the 1–2 mm caterpillar size range.

Space for calculation

_____ %   1

(d) Describe the relationship between beak size index and caterpillar size range eaten.

_____

_____   1

(e) The average beak length of Great Tits is 13 mm.

Calculate their average beak depth.

Space for calculation

_____ mm   1

(f) From the information given, describe all of the ways by which interspecific competition for caterpillars is reduced between the following pairs of bird species.

(i) Blue Tits and Great Tits _____

_____

_____   1

(ii) Blue Tits and Marsh Tits _____

_____

_____   1

(g) Each bird species must forage economically.

Explain what is meant by this statement in terms of energy gain and loss.

_____

_____

_____   1

Marks

9. The ancestor species of the modern tomato produces small fruits but can grow in soils with low nitrogen levels.

   The modern species of tomato has been selectively bred to produce large fruit but requires soil rich in nitrogen to grow well.

   (a)   (i)   Give **one** reason why nitrogen is important for plant growth.

   _____

   _____   1

         (ii)   Give **one** symptom of nitrogen deficiency in plants.

   _____   1

   (b)   The diagram represents steps in a technique used by tomato breeders to combine characteristics of these two species.

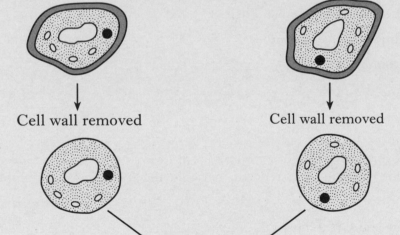

Leaf cell from ancestor plant   Leaf cell from selectively bred plant

Cell wall removed   Cell wall removed

Remaining cell material combines

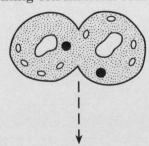

New species of tomato plant

DO NOT WRITE IN THIS MARGIN

Marks

9. (*b*) (continued)

(i) Give the name of the technique shown in the diagram.

____ **1**

(ii) Name the enzyme that is used to remove the cell walls from the leaf cells.

____ **1**

(iii) Explain why cultivation of the new species of tomato could lead to a reduction in the use of nitrogen fertiliser.

____

____

____ **1**

[Turn over

Marks

10. An investigation was carried out to compare the growth of *Escherichia coli* (*E. coli*) bacteria in different nutrient solutions.

E. coli were grown in a glucose solution for 24 hours. Two $50\,cm^3$ samples were transferred to two identical containers each with different sterile nutrient solutions as shown in the table.

Container	Nutrient solution
X	0·5 mM glucose
Y	0·5 mM lactose

One container is shown in the diagram.

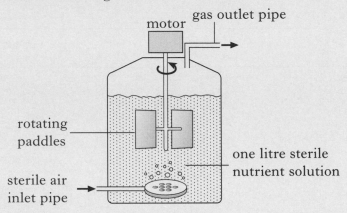

The pH and temperature were kept constant.

Every 30 minutes, a $2\,cm^3$ sample was taken from each container.

An instrument was used to measure the number of bacteria present. The higher the instrument reading, the more bacteria.

(a) (i) Suggest a reason for having rotating paddles.

_____

_____ 1

    (ii) Explain why a gas outlet pipe is needed in the apparatus.

_____

_____

_____ 1

(b) Identify **two** variables not already mentioned that would have to be controlled in both containers to make the procedure valid.

1 _____ 1

2 _____ 1

Marks

10. (continued)

(c) The results of the investigation are shown in the graph.

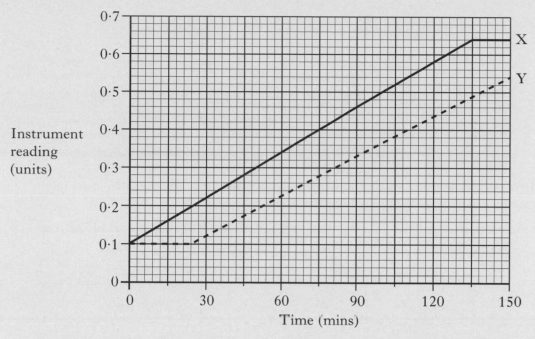

E. coli only produces the enzyme to metabolise lactose when there is no glucose available to the cells.

(i) Use information from the graph to state the time taken for the bacterial cells to produce the enzyme needed to metabolise lactose.

Justify your answer.

Time _____ minutes

Justification _____

_____

_____   **1**

(ii) Explain how lactose acts as an inducer of this enzyme.

_____

_____

_____   **2**

(iii) Predict the instrument reading at 150 minutes if a third container had been used with nutrient solution containing 0·25 mM glucose.

_____ units   **1**

Marks

11. In seed pods of garden pea plants, smooth shape (T) is dominant to constricted shape (t) and green (G) is dominant to yellow (g).

The genes are not linked.

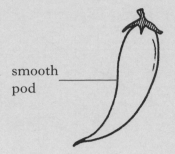

smooth pod

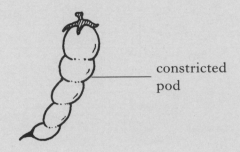

constricted pod

(*a*) In an investigation, a pea plant heterozygous for both smooth and green pods was crossed with a pea plant with constricted, yellow pods.

Complete the table to give the genotypes of the parent plants and **all** of their possible gametes.

Phenotype of parent	Genotype of parent	Genotype(s) of gamete(s)
Smooth green pod		
Constricted yellow pod		

1

1

(*b*) In a second investigation, two pea plants heterozygous for both seed shape and seed colour were crossed. This produced 112 offspring.

   (i) Calculate the **expected** number of offspring that would have yellow pods.

Space for calculation

_____ 1

The actual phenotypes obtained did not occur in the expected numbers.

   (ii) Suggest **two** reasons why the **actual** numbers observed may differ from the **expected** numbers in a dihybrid cross.

1 _____

_____ 1

2 _____

_____ 1

Marks

12. A potato tuber is a swollen underground stem.

Two similar tubers, each with one apical bud and four lateral buds were treated as shown in the diagram. Their appearance after being kept in the dark for three weeks is also shown.

Apical buds produce the plant growth substance IAA.

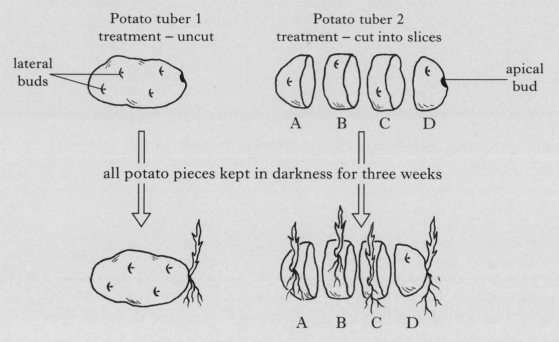

(*a*) Name the effect of IAA that is shown by potato tuber 1 after three weeks.

_____   **1**

(*b*) Explain why the lateral buds on slices A, B and C produced shoots.

_____

_____

_____   **1**

(*c*) Potato tuber 1 was then exposed to light from one direction for a further three days. The shoot showed phototropism.

Explain the role of IAA in phototropism.

_____

_____

_____   **2**

(*d*) Potatoes are long day plants. In terms of changing photoperiod, describe what is meant by "long day plant".

_____

_____   **1**

Marks

13. The table shows the mass of salmon caught in coastal waters around Scotland.

Year	Scottish catch (tonnes)
1973	1300
1980	1200
1988	900
1998	200

(a) By how many times was the 1973 catch greater than the 1998 catch?

Space for calculation

_____ times   **1**

(b) Scientists monitoring the salmon have suggested the following four possible factors for the decline in numbers.

- Predation by seals

- Food shortage

- Rising sea temperature

- Infection by sea louse parasites

Underline the factor(s) that would have had a density-independent effect on the salmon population.   **1**

(c) Populations of North Atlantic salmon are monitored because they are a food species.

Give **one** other reason for monitoring animal populations.

_____

_____   **1**

Marks

14. The table shows how two structures in mammalian skin respond to a drop in the surrounding air temperature from 20 °C to 5 °C.

Structure	Air temperature	
	20 °C	5 °C
hair erector muscles	relaxed	contracted
blood vessels	dilated	constricted

(a) (i) Name the temperature monitoring centre in the body of a mammal.

_____ 1

(ii) State how messages are sent from the temperature monitoring centre to the skin.

_____ 1

(b) Explain the advantage to the organism of constriction of skin blood vessels when the air temperature drops from 20 °C to 5 °C.

_____

_____

_____ 1

(c) Give the term that describes an animal that obtains most of its body heat from its own metabolism.

_____ 1

[Turn over for Section C on *Page thirty-four*

SECTION C

Both questions in this section should be attempted.

Note that each question contains a choice.

Questions 1 and 2 should be attempted on the blank pages which follow.

Supplementary sheets, if required, may be obtained from the invigilator.

All answers must be written clearly and legibly in ink.

Labelled diagrams may be used where appropriate.

Marks

1. Answer **either** A **or** B.

   A. Give an account of transpiration under the following headings:

   (i)   the effect of environmental factors on transpiration rate;   **5**

   (ii) adaptations of xerophyte plants that reduce the transpiration rate.   **5**

   OR   **(10)**

   B. Give an account of how animals and plants cope with dangers under the following headings:

   (i)   behavioural defence mechanisms in animals;   **5**

   (ii) cellular and structural defence mechanisms in plants.   **5**

   (10)

In question 2, ONE mark is available for coherence and ONE mark is available for relevance.

2. Answer **either** A **or** B.

   A. Give an account of the principle of negative feedback with reference to the maintenance of blood sugar levels.   **(10)**

   OR

   B. Give an account of the role of the pituitary gland in controlling normal growth and development and describe the effects of named drugs on fetal development.   **(10)**

[END OF QUESTION PAPER]

SPACE FOR ANSWERS

SPACE FOR ANSWERS

SPACE FOR ANSWERS

SPACE FOR ANSWERS

SPACE FOR ANSWERS

SPACE FOR ANSWERS

SPACE FOR ANSWERS

SPACE FOR ANSWERS

SPACE FOR ANSWERS

SPACE FOR ANSWERS

ADDITIONAL GRAPH PAPER FOR QUESTION 1(*b*)(i)

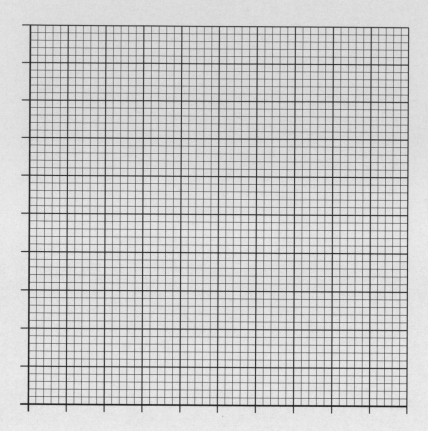

[BLANK PAGE]

FOR OFFICIAL USE

Total for
Sections
B and C

X007/301

NATIONAL
QUALIFICATIONS
2007

MONDAY, 21 MAY
1.00 PM – 3.30 PM

BIOLOGY
HIGHER

Fill in these boxes and read what is printed below.

Full name of centre

Town

Forename(s)

Surname

Date of birth
Day Month Year   Scottish candidate number   Number of seat

SECTION A—Questions 1–30 (30 marks)

Instructions for completion of Section A are given on page two.

For this section of the examination you must use an **HB pencil**.

SECTIONS B AND C (100 marks)

1   (a) All questions should be attempted.

(b) It should be noted that in **Section C** questions 1 and 2 each contain a choice.

2   The questions may be answered in any order but all answers are to be written in the spaces provided in this answer book, **and must be written clearly and legibly in ink**.

3   Additional space for answers will be found at the end of the book. If further space is required, supplementary sheets may be obtained from the invigilator and should be inserted inside the **front** cover of this book.

4   The numbers of questions must be clearly inserted with any answers written in the additional space.

5   Rough work, if any should be necessary, should be written in this book and then scored through when the fair copy has been written. If further space is required a supplementary sheet for rough work may be obtained from the invigilator.

6   Before leaving the examination room you must give this book to the invigilator. If you do not, you may lose all the marks for this paper.

SCOTTISH
QUALIFICATIONS
AUTHORITY

Read carefully

1 Check that the answer sheet provided is for **Biology Higher (Section A)**.

2 For this section of the examination you must use an **HB pencil**, and where necessary, an eraser.

3 Check that the answer sheet you have been given has **your name**, **date of birth**, **SCN** (Scottish Candidate Number) and **Centre Name** printed on it.

Do not change any of these details.

4 If any of this information is wrong, tell the Invigilator immediately.

5 If this information is correct, **print** your name and seat number in the boxes provided.

6 The answer to each question is **either** A, B, C or D. Decide what your answer is, then, using your pencil, put a horizontal line in the space provided (see sample question below).

7 There is **only one correct** answer to each question.

8 Any rough working should be done on the question paper or the rough working sheet, **not** on your answer sheet.

9 At the end of the exam, put the **answer sheet for Section A inside the front cover of this answer book**.

Sample Question

The apparatus used to determine the energy stored in a foodstuff is a

A calorimeter

B respirometer

C klinostat

D gas burette.

The correct answer is **A**—calorimeter. The answer **A** has been clearly marked in **pencil** with a horizontal line (see below).

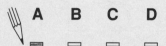

Changing an answer

If you decide to change your answer, carefully erase your first answer and using your pencil fill in the answer you want. The answer below has been changed to **D**.

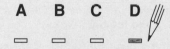

SECTION A

All questions in this section should be attempted.

Answers should be given on the separate answer sheet provided.

1. Which line in the table identifies correctly the two cell structures shown in the diagram?

	X	Y
A	Golgi body	Vesicle
B	Golgi body	Ribosome
C	Endoplasmic reticulum	Vesicle
D	Endoplasmic reticulum	Ribosome

2. The phospholipid molecules in a cell membrane allow the

   A free passage of glucose molecules

   B self-recognition of cells

   C active transport of ions

   D membrane to be fluid.

3. Red blood cells have a solute concentration of around 0·9%.

   Which of the following statements correctly describes the fate of these cells when immersed in a 1% salt solution?

   A The cells will burst.

   B The cells will shrink.

   C The cells will expand but not burst.

   D The cells will remain unaffected.

4. Which graph best illustrates the effect of increasing temperature on the rate of active uptake of ions by roots?

A
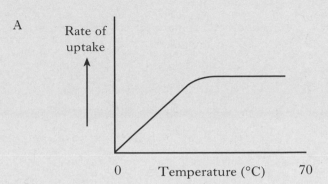

B
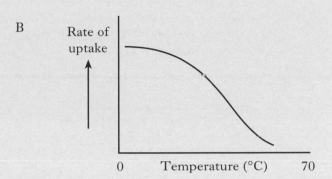

C
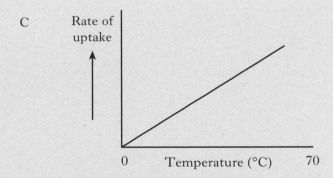

D
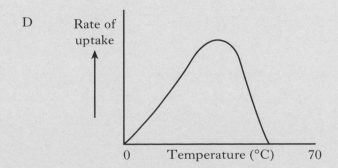

5. Which substances must be provided by host cells for the synthesis of viruses?

    A Proteins and nucleotides

    B Amino acids and DNA

    C Proteins and DNA

    D Amino acids and nucleotides

6. The action spectrum of photosynthesis is a measure of the ability of plants to

    A absorb all wavelengths of light

    B absorb light of different intensities

    C use light to build up foods

    D use light of different wavelengths for synthesis.

7. The diagram shows DNA during replication. Base H represents thymine and base M represents guanine. Which letters represent the base cytosine?

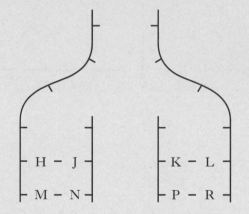

    A J and K

    B J and L

    C N and P

    D N and R

8. The graph below shows the effect of light intensity on the rate of photosynthesis at different temperatures.

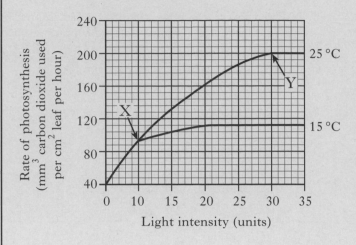

Which of the following conclusions can be made from the above data?

    A Only at light intensities greater than 20 units does temperature affect the rate of photosynthesis.

    B At point Y, the rate of photosynthesis is limited by the light intensity.

    C Temperature has little effect on the rate of photosynthesis at low light intensities.

    D At point X, temperature limits the rate of photosynthesis.

9. The cell structures shown below have been magnified ten thousand times.

Mitochondrion   Chloroplast

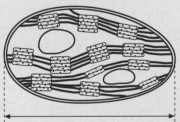

0·04 micrometres

0·065 micrometres

Expressed as a simple whole number ratio, the length of the mitochondrion compared to that of the chloroplast is

    A   8 : 13

    B 13 : 8

    C 40 : 65

    D 65 : 40.

10. A section of a DNA molecule contains 300 bases. Of these bases, 90 are adenine. How many cytosine bases would this section of DNA contain?

   A   60

   B   90

   C   120

   D   180

11. What information can be derived from the recombination frequencies of linked genes?

   A   The mutation rate of the genes

   B   The order and location of genes on a chromosome

   C   Whether genes are recessive or dominant

   D   The genotype for a particular characteristic

12. The diagram shows a family tree for a family with a history of red-green colour deficiency.

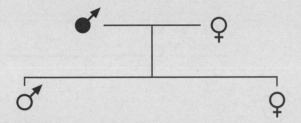

 Unaffected male

 Affected male

 Unaffected female

 Affected female

The allele for red-green colour deficiency is sex-linked.

Which of the following statements is true?

   A   Only the son is a carrier.

   B   Only the daughter is a carrier.

   C   Both son and daughter are carriers.

   D   Neither son nor daughter is a carrier.

13. Apple crop yields have been increased by plant breeders selecting for

   A   disease resistance

   B   flavour

   C   resistance to bruising

   D   sugar content.

14. Human insulin can be produced by the bacterium *E. coli* using the following steps.

   1   Culture large quantities of *E. coli* in vats of nutrients.

   2   Insert human insulin gene into *E. coli* plasmid DNA.

   3   Cut insulin gene from human chromosome using enzymes.

   4   Extract insulin from culture vats.

   The correct order for these steps is

   A   3, 2, 1, 4

   B   3, 1, 2, 4

   C   1, 4, 3, 2

   D   1, 2, 3, 4.

15. In a desert mammal, which of the following is a physiological adaptation which helps to conserve water?

   A   Nocturnal foraging

   B   Breathing humid air in a burrow

   C   Having few sweat glands

   D   Remaining underground by day

16. The following factors affect the transpiration rate in a plant.

   1   increasing wind speed

   2   decreasing humidity

   3   rising air pressure

   4   falling temperature

   Which two of these factors would cause an increase in transpiration rate?

   A   1 and 2

   B   1 and 3

   C   2 and 4

   D   3 and 4

[Turn over

17. When first exposed to a harmless stimulus, a group of animals responded by showing avoidance behaviour. When the stimulus was repeated the animals became habituated to it.

What change in response would have shown that habituation was taking place?

A   An increase in the length of the response

B   A decrease in the time taken to respond

C   An increase in response to other stimuli

D   A decrease in the percentage of animals responding

18. In tomato plants, the allele for curled leaves is dominant over the allele for straight leaves. The allele for hairy stems is dominant over the allele for hairless stems. The genes for curliness and hairiness are located on different chromosomes.

If plants heterozygous for both characteristics were crossed, what ratio of phenotypes would be expected in the offspring?

A   All curly and hairy

B   3 curly and hairy: 1 straight and hairless

C   9 curly and hairy: 3 curly and hairless:
    3 straight and hairy: 1 straight and hairless

D   1 curly and hairy: 1 curly and hairless:
    1 straight and hairy: 1 straight and hairless

19. The bar graph below shows changes in the DNA content per cell during stages of meiosis.

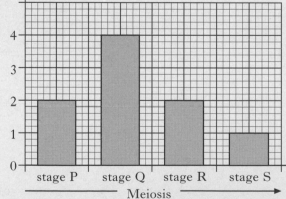

When do the homologous pairs of chromosomes separate?

A   Before the start of stage P

B   Between stages P and Q

C   Between stages Q and R

D   Between stages R and S

20. The genes for two different characteristics are located on separate chromosomes.

In a cross between individuals with the genotypes AaBb and aabb, what is the chance of any one of the offspring having the genotype aabb?

A   0

B   1 in 2

C   1 in 4

D   1 in 8

21. Which line in the table identifies correctly the main source of body heat and the method of controlling body temperature in an endotherm?

	Main source of body heat	Principal method of controlling body temperature
A	Respiration	Behavioural
B	Respiration	Physiological
C	Absorbed from environment	Behavioural
D	Absorbed from environment	Physiological

22. The following four statements relate to meristems.

1   Some provide cells for increase in diameter in stems

2   Some produce growth substances

3   They are found in all growing organisms

4   Their cells undergo division by meiosis

Which of the above statements are true?

A   1 and 2 only

B   1 and 3 only

C   2 and 3 only

D   2 and 4 only

23. The diagram below shows a transverse section of a woody stem.

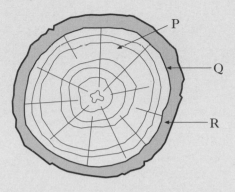

Which line of the table identifies correctly the tissues P, Q and R?

	P	Q	R
A	cambium	xylem	phloem
B	phloem	xylem	cambium
C	xylem	phloem	cambium
D	xylem	cambium	phloem

24. The graph below shows changes which occur in the masses of protein, fat and carbohydrate in a girl's body during seven weeks of starvation.

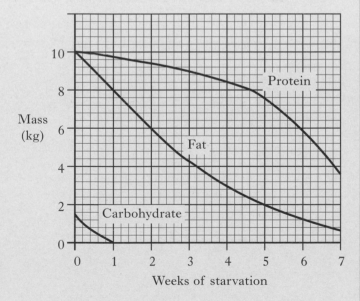

The girl weighs 60 kg at the start. Predict her weight after two weeks without food.

A   43 kg

B   50 kg

C   54 kg

D   57 kg

25. Which of the following is triggered by the hypothalamus in response to an increase in the temperature of the body?

A   Contraction of the hair erector muscles and vasodilation of the skin capillaries

B   Relaxation of the hair erector muscles and vasodilation of the skin capillaries

C   Contraction of the hair erector muscles and vasoconstriction of the skin capillaries

D   Relaxation of the hair erector muscles and vasoconstriction of the skin capillaries

26. Plants require macro-elements for the synthesis of various compounds. Identify which macro-elements are required for synthesis of the compounds shown in the table below.

	Chlorophyll	Protein	ATP
A	phosphorus	magnesium	nitrogen
B	phosphorus	nitrogen	magnesium
C	magnesium	nitrogen	phosphorus
D	magnesium	phosphorus	nitrogen

27. The following are events occurring during germination in barley.

1   The embryo produces gibberellic acid (GA)

2   α-amylase is produced

3   Gibberellic acid (GA) passes to the aleurone layer

4   α-amylase converts starch to maltose

5   Maltose is used by the embryo

Which of the following indicates the correct sequence of events?

A   1 2 4 5 3

B   1 3 2 4 5

C   3 2 1 4 5

D   5 1 3 2 4

[Turn over

28. The table shows the masses of various substances in the glomerular filtrate and in the urine over a period of 24 hours.

Which of the substances has the smallest percentage of reabsorption from the glomerular filtrate?

	Substance	Mass in glomerular filtrate (g)	Mass in urine (g)
A	Sodium	600·0	6·0
B	Potassium	35·0	2·0
C	Uric acid	8·5	0·8
D	Calcium	5·0	0·2

29. An investigation was carried out into the effect of indole acetic acid (IAA) concentration on the shoot growth of two species of plant. The graph below shows a summary of the results.

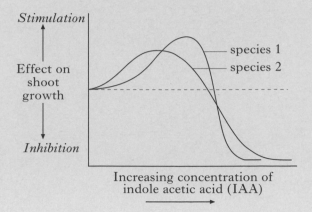

Which one of the following conclusions is justified?

A Species 1 shows its maximum stimulation at a lower IAA concentration than species 2.

B Species 2 is more inhibited by the highest concentrations of IAA than species 1.

C Species 2 is stimulated over a greater range of IAA concentrations than species 1.

D Species 1 is stimulated by some IAA concentrations which inhibit species 2.

30. An enzyme and its substrate were incubated with various concentrations of either copper or magnesium salts.

The time taken for the complete breakdown of the substrate was measured.

The results are given in the table.

Salt Concentration (M)	Time needed to break down substrate (s)	
	Copper salts	Magnesium salts
0	39	39
1×10^{-8}	42	21
1×10^{-6}	380	49
1×10^{-4}	1480	286

(*increasing concentration* ↓)

From the data, it may be deduced that

A high concentrations of copper salts promote the activity of the enzyme

B high concentrations of copper salts inhibit the activity of the enzyme

C low concentrations of magnesium salts inhibit the activity of the enzyme

D high concentrations of magnesium salts promote the activity of the enzyme.

Candidates are reminded that the answer sheet MUST be returned INSIDE the front cover of this answer book.

[Turn over for Section B on *Page ten*

SECTION B

All questions in this section should be attempted.

All answers must be written clearly and legibly in ink.

Marks

1. (*a*) The diagram contains information about light striking a leaf.

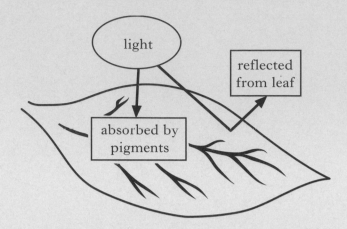

   (i) Apart from being absorbed or reflected, what can happen to light which strikes a leaf?

   _____   1

   (ii) Pigments that absorb light are found within leaf cells.

   State the exact location of these pigments.

   _____   1

   (*b*) The diagram below shows part of the light dependent stage of photosynthesis.

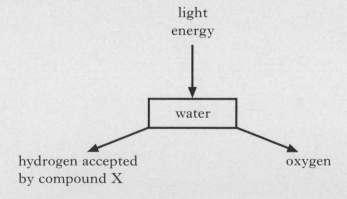

   (i) Name this part of the light dependent stage.

   _____   1

   (ii) Name compound X.

   _____   1

Marks

1. **(continued)**

    (*c*) The following sentences describe events in the carbon fixation stage of photosynthesis.

    Underline one alternative in each pair to make the sentences correct.

    The $\begin{Bmatrix} \text{three} \\ \text{five} \end{Bmatrix}$ carbon compound ribulose bisphosphate (RuBP) accepts

    $\begin{Bmatrix} \text{carbon dioxide} \\ \text{hydrogen} \end{Bmatrix}$.

    $\begin{Bmatrix} \text{Carbon dioxide} \\ \text{Hydrogen} \end{Bmatrix}$ is accepted by the $\begin{Bmatrix} \text{three} \\ \text{five} \end{Bmatrix}$ carbon compound

    glycerate phosphate (GP).

    2

    [Turn over

Marks

2. The diagram shows apparatus set up to investigate the rate of respiration in an earthworm. After 10 minutes at 20 °C the level of liquid in the capillary tube had changed as shown.

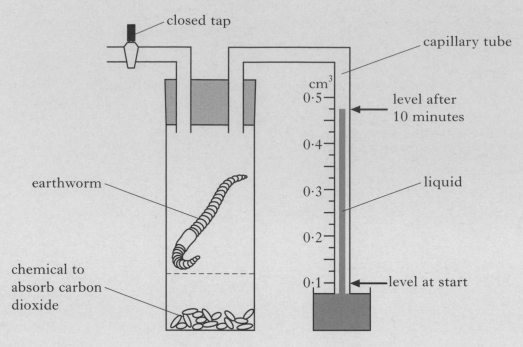

closed tap

capillary tube

cm³

0·5 — level after 10 minutes

0·4

earthworm

0·3 — liquid

0·2

chemical to absorb carbon dioxide

0·1 — level at start

(a) (i) What volume of oxygen is used by the earthworm during the 10 minute period?

_____ cm³   **1**

(ii) Describe a suitable control for this experiment.

_____

_____   **1**

(b) In a second experiment, a worm of 5 grams used 0·5 cm³ of oxygen in 10 minutes.

Calculate its rate of respiration in cm³ per minute per gram of worm.

Space for calculation

_____ cm³ per minute per gram of worm   **1**

[Turn over for Question 3 on *Page fourteen*

Marks

3. (*a*) Samples of carrot tissue were immersed in a hypotonic solution at two different temperatures for 5 hours. The mass of the tissue samples was measured every hour and the percentage change in mass calculated.

The results are shown on the graph.

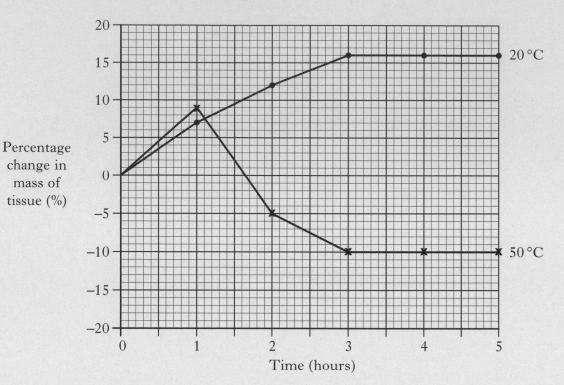

(i) Explain the results obtained at 20 °C from 0 to 3 hours and from 3 hours to 5 hours.

0 to 3 hours _____

_____   **1**

3 to 5 hours _____

_____   **1**

(ii) Explain the change in mass of the carrot tissue between 1 and 3 hours at 50 °C.

_____

_____

_____   **2**

Marks

3. **(continued)**

(*b*) The chart shows the concentration of ions within a unicellular organism and in the sea water surrounding it.

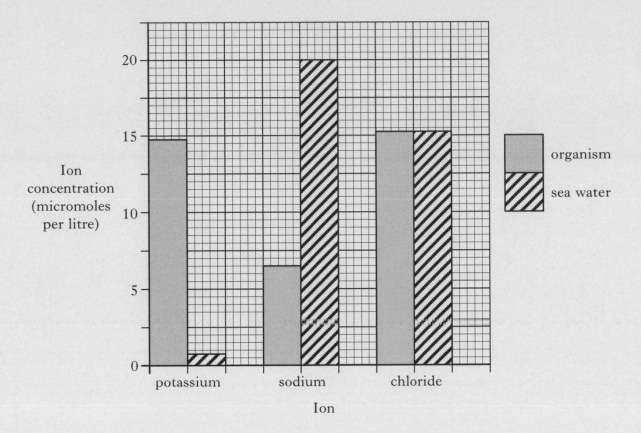

(i) From the information given, identify the ion which appears to move between the organism and the sea water by diffusion.

Justify your choice.

Ion _____

Justification _____

_____

   1

(ii) When oxygen was bubbled through a tank of sea water containing these organisms, the potassium ion concentration within the organisms increased.

Explain this effect.

_____

_____

_____

   2

[Turn over

DO NOT WRITE IN THIS MARGIN

Marks

4. The diagram shows events occurring during the synthesis of a protein that is secreted from a cell.

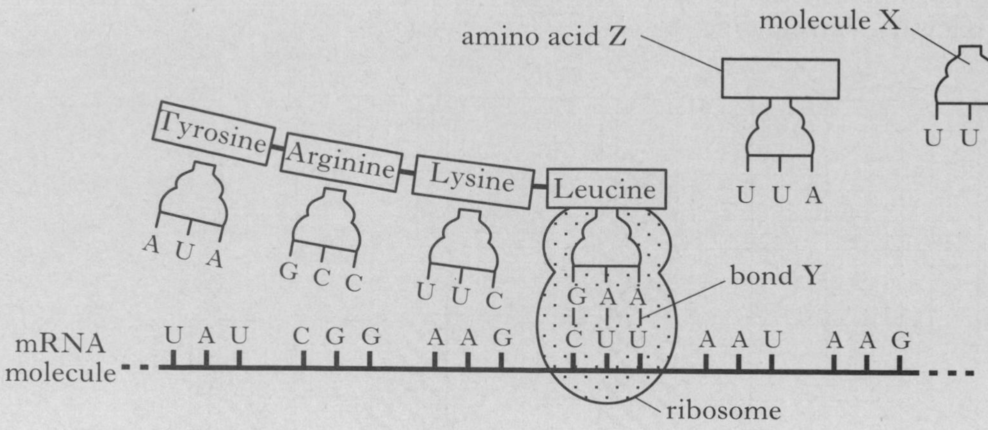

(*a*) (i) Name molecule X. ______ **1**

(ii) Name bond Y. ______ **1**

(*b*) What name is given to a group of three bases on mRNA that codes for an amino acid?

______ **1**

(*c*) Give the sequence of DNA bases that codes for amino acid Z.

______ **1**

(*d*) Describe the roles of the endoplasmic reticulum and the Golgi apparatus between the synthesis of the protein and its release from the cell.

Endoplasmic reticulum ______

______ **1**

Golgi apparatus ______

______ **1**

(*e*) The table contains some information about the structure and function of proteins.

Add information to the boxes to complete the table.

Protein	*Structure (Globular or Fibrous)*	*Function*
Cellulase		
Collagen		Structural protein in skin

2

Marks

5. (*a*) The graph shows the relationship between plant species diversity in grassland and grazing intensity by herbivores.

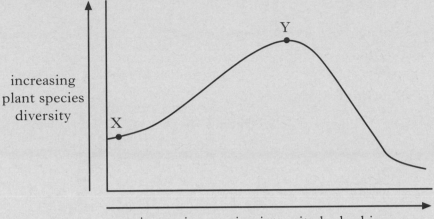

increasing grazing intensity by herbivores

(i) Explain the effect on plant species diversity of increased grazing intensity by herbivores between X and Y on the graph.

_____

_____

_____   2

(ii) What evidence is there that grassland contains plant species tolerant of grazing?

_____

_____   1

(*b*) State **one** feature of some plant species that allows them to tolerate grazing by herbivores.

_____

_____   1

[Turn over

6. Norway Spruce (*Picea abies*) is an evergreen species of tree with needle-like leaves, found in regions with extremely cold winters.

The rate of photosynthesis of the species is at its maximum during spring then decreases from June to December.

In an investigation, a sample of one-year-old seedlings was collected in each month from June to December.

For each sample of seedlings, the following measurements were made and averages calculated.

- Dry mass of whole seedlings
- Dry mass of roots only
- Starch content in needles
- Sugar content in needles

The results are shown in **Graphs 1** and **2**.

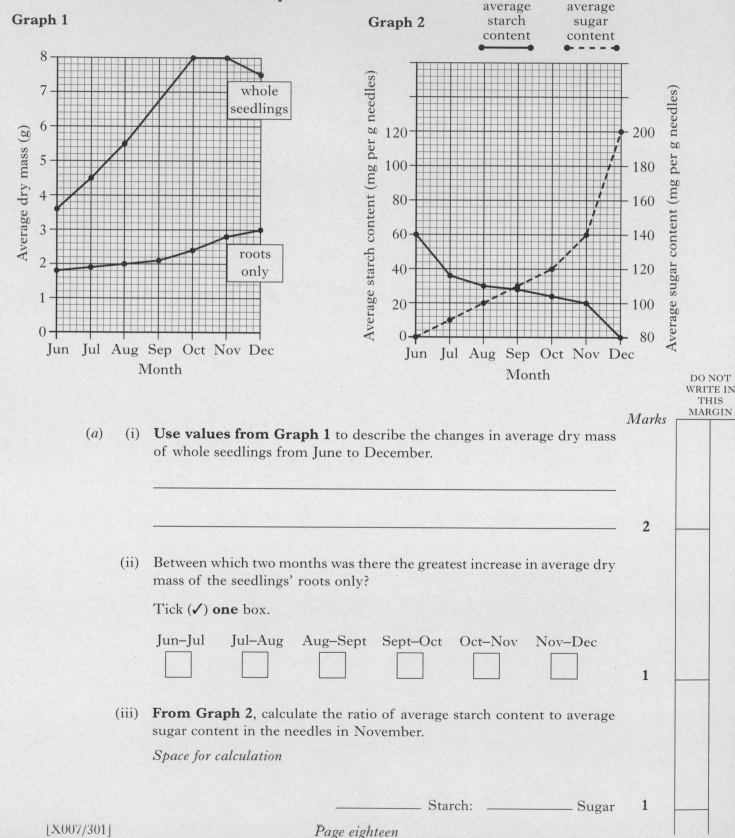

Graph 1

Graph 2

Marks

(a) (i) **Use values from Graph 1** to describe the changes in average dry mass of whole seedlings from June to December.

_____

_____   **2**

(ii) Between which two months was there the greatest increase in average dry mass of the seedlings' roots only?

Tick (✓) **one** box.

Jun–Jul   Jul–Aug   Aug–Sept   Sept–Oct   Oct–Nov   Nov–Dec

☐   ☐   ☐   ☐   ☐   ☐   **1**

(iii) **From Graph 2**, calculate the ratio of average starch content to average sugar content in the needles in November.

Space for calculation

_____ Starch: _____ Sugar   **1**

Marks

6. **(a)** **(continued)**

    (iv) **From Graph 2**, calculate the percentage decrease in average starch content in the needles between June and October.

    Space for calculation

    _____ % decrease   **1**

(b) Explain the decrease in average starch content in the needles between June and December.

_____

_____   **1**

(c) Raffinose is a sugar that prevents frost damage to needles.

The table shows the raffinose content of needles from the seedling samples.

Month	Raffinose content (mg per g of needles)
June	0
July	1
August	2
September	3
October	9
November	30
December	50

    (i) What evidence is there that raffinose is not the only sugar present in the needles of Norway Spruce?

_____

_____   **1**

    (ii) Suggest how the changing raffinose content of needles from June to December is of survival value to Norway Spruce.

_____

_____   **1**

[Turn over

Marks

7. The diagram shows stages in meiosis during which a mutation occurred and the effect of the mutation on the gametes produced.

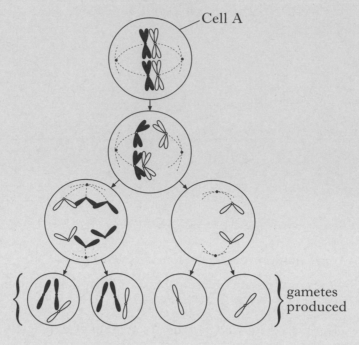

Cell A

gametes
produced

(*a*) (i) What name is given to cells such as cell A, that undergo meiosis?

_____   1

(ii) Cell A contains two pairs of homologous chromosomes.

Apart from size and shape, state **one** similarity between homologous chromosomes.

_____   1

(iii) This mutation has resulted in changes to the chromosome numbers in the gametes.

Name this type of mutation.

_____   1

(iv) State whether the mutation has occurred in the first or second meiotic division and justify your choice.

Meiotic division _____

Justification _____

_____   1

(v) State the expected haploid number of chromosomes in the gametes produced if this mutation **had not occurred**.

_____   1

7. (continued)

Marks

(b) The diagram represents the sequence of bases on part of one strand of a DNA molecule.

Part of DNA molecule   T   G   A   A   C   T   G

The effects of two different gene mutations on the strand of DNA are shown below.

Gene mutation 1   T   T   G   A   A   C   T   G

Gene mutation 2   T   G   A   C   C   T   G

Complete the table by naming the type of gene mutation that has occurred in each case.

Gene Mutation	Name
1	
2	

2

[Turn over

Marks

8. The table shows information about three species of oak tree that have evolved from a common ancestor.

	Oak Species		
	Sessile Oak	Kermes Oak	Northern Red Oak
Leaf Shape	Rounded lobes	Sharp spines	Lobes with sharp spines
Growing Conditions	Mild and damp	Hot and dry	Cool and dry

(*a*) (i) The Oak species have evolved in ecological isolation.

State the importance of isolating mechanisms in the evolution of new species.

_____   1

(ii) Use the information to explain how the evolution of the Oak species illustrates adaptive radiation.

_____

_____   2

(*b*) The Kermes Oak grows to a maximum height of one metre.

Explain the benefit to this species of having leaves with sharp spines.

_____

_____   1

(*c*) To maintain genetic diversity, species must be conserved.

State **two** ways in which species can be conserved.

1 _____

2 _____   1

Marks

9. (*a*) The graph shows the body mass of a human male from birth until 22 years of age.

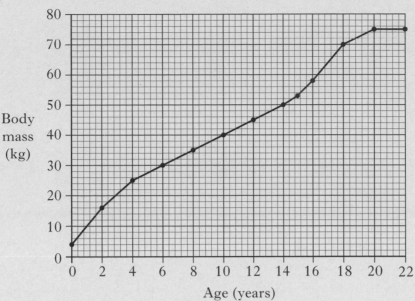

(i) Calculate the average yearly increase in body mass between age 6 and age 14.

Space for calculation

_____ kg   **1**

(ii) Explain how the activity of the pituitary gland could account for the growth pattern between 15 and 22 years of age shown on the graph.

_____

_____

_____   **2**

(*b*) The diagram shows the role of the pituitary gland in the secretion of a hormone from the thyroid gland.

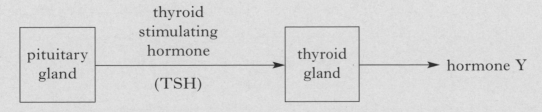

Name hormone Y and describe its role in the control of growth and development.

Hormone Y_____   **1**

Role _____   **1**

Marks

10. An experiment was carried out to investigate the effect of gibberellic acid (GA) on the growth of dwarf pea plants. GA can be absorbed by leaves.

Six identical pea plants were placed in pots containing 100 g of soil. The leaves of each were sprayed with an equal volume of water containing a different mass of GA. The soil in each pot received 20 cm³ water each day and the plants were continuously exposed to equal light intensity from above.

After seven days the stem height of each plant was measured and the percentage increase in stem height calculated.

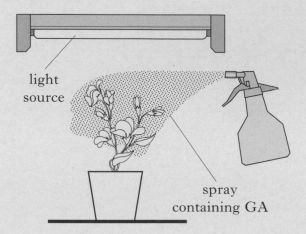

(*a*) (i) Identify **two** variables, not already mentioned, that should have been controlled to ensure the experimental procedure was valid.

Variable 1 _____   1

Variable 2 _____   1

(ii) State **one** way in which the experimental procedure could be improved to increase the reliability of the results.

_____

_____   1

(*b*) The results of the experiment are shown in the table.

Mass of GA applied (micrograms)	Percentage increase in stem height (%)
0·01	90
0·03	120
0·05	160
0·08	240
0·10	320
0·11	350

Marks

10. **(*b*)** **(continued)**

(i) On the grid provided, complete the line graph to show the percentage increase in stem height against the mass of GA applied.

Use an appropriate scale to fill most of the grid.

(Additional graph paper, if required, will be found on page 36.)

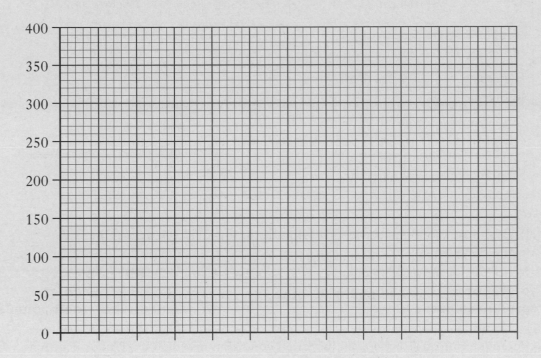

2

(ii) Another pea plant was treated in the same way, using a water spray containing 0·12 micrograms of GA. Predict the percentage increase in stem height of this plant after seven days.

_____ percentage increase   1

(*c*) Explain why the method of application of GA could lead to errors in the results.

_____

_____   1

[Turn over

Marks

11. (*a*) The bacterium *Escherichia coli* can control its lactose metabolism.

Complete **all** boxes in the table to show whether each statement is true (T) or false (F) if lactose is present or absent in the medium in which *E. coli* is growing.

Statement	*Lactose present*	*Lactose absent*
Regulator gene produces the repressor molecule		
Repressor molecule binds to inducer		
Repressor molecule binds to operator		
Structural gene switched on	T	F

2

(*b*) Part of a metabolic pathway involving the amino acid phenylalanine is shown in the diagram.

digestion of protein in diet ⟶ **phenylalanine** —enzyme A⟶ **tyrosine** —enzyme B⟶ **other compounds**

Phenylketonuria (PKU) is an inherited condition in which enzyme A is either absent or does not function.

(i) Predict the effect on the concentrations of phenylalanine and tyrosine if enzyme A is absent.

Phenylalanine _____ 1

Tyrosine _____ 1

(ii) PKU is caused by a mutation of the gene that codes for enzyme A.

Explain how a mutation of a gene can cause the production of an altered enzyme.

_____

_____

_____ 2

Marks

12. An experiment was set up to investigate the effect of photoperiod on flowering in *Chrysanthemum* plants. Four plants A, B, C and D were exposed to different periods of light and dark in 24 hours. This was repeated every day for several weeks and the effects on flowering noted.

The periods of light and dark and their effects on flowering are shown in the diagram.

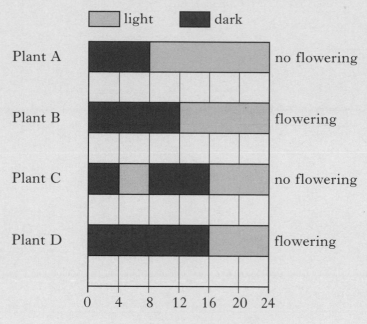

Periods of light and dark (hours)

(a) From the information given, identify the conditions required for flowering in *Chrysanthemum* plants. Justify your answer.

Conditions _____

Justification _____

_____ 1

(b) Flowering in response to photoperiod ensures plants within a population flower at the same time. Explain how this enables genetic variation to be maintained.

_____

_____ 1

(c) Mammals also show photoperiodism.

Describe how one type of mammal behaviour can be affected by photoperiod.

_____

_____ 1

[Turn over

Marks

13. The homeostatic control of blood glucose concentration carried out by the human liver is shown on the diagram.

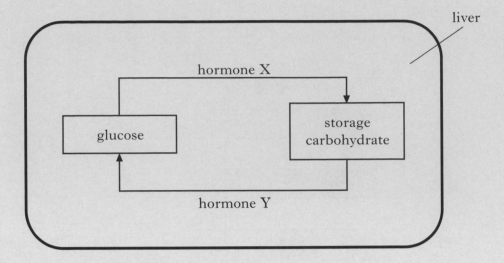

(*a*)   Name the storage carbohydrate found in the liver.

_____   1

(*b*)   (i)   Name hormones X and Y.

Hormone X _____

Hormone Y _____   1

(ii)   Name the organ that produces hormones X and Y.

_____   1

(iii) Explain how negative feedback is involved in the homeostatic control of blood glucose concentration.

_____

_____

_____   2

Marks

14. (*a*) The diagram shows some plant communities present at various time intervals on farmland cleared of vegetation by a fire.

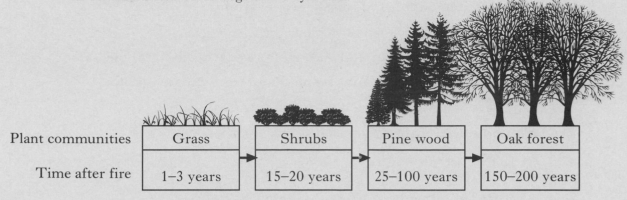

Plant communities	Grass	Shrubs	Pine wood	Oak forest
Time after fire	1–3 years	15–20 years	25–100 years	150–200 years

(i) State the term used to describe this sequence of plant communities.

_____ **1**

(ii) Give a reason to explain why the shrub community is able to replace the grass community after 15 years.

_____ **1**

(iii) Oak forest is the climax community in this sequence.
Describe a feature of a climax community.

_____ **1**

(*b*) The grid shows factors that can influence population change.

A competition	B predation	C rainfall
D disease	E temperature	F food supply

(i) Use **all** the letters from the grid to complete the table to show which factors are density dependent and which are density independent.

Density dependent	*Density independent*

2

(ii) **Underline** one alternative in each pair to make the sentences correct.

As population density $\begin{Bmatrix} \text{increases,} \\ \text{decreases,} \end{Bmatrix}$ the effect of density dependent factors increases.

As a result, the population density then $\begin{Bmatrix} \text{increases} \\ \text{decreases} \end{Bmatrix}$. **1**

[Turn over for Section C on *Page thirty*

SECTION C

Both questions in this section should be attempted.

Note that each question contains a choice.

Questions 1 and 2 should be attempted on the blank pages which follow.

Supplementary sheets, if required, may be obtained from the invigilator.

All answers must be written clearly and legibly in ink.

Labelled diagrams may be used where appropriate.

Marks

1. Answer **either** A **or** B.

    A. Give an account of respiration under the following headings:

    (i)   glycolysis;   **5**

    (ii) the Krebs (Citric acid) cycle.   **5**

                                                                 (10)

    OR

    B. Give an account of cellular defence mechanisms in animals under the following headings:

    (i)   phagocytosis;   **4**

    (ii) antibody production and tissue rejection.   **6**

                                                                 (10)

In question 2, ONE mark is available for coherence and ONE mark is available for relevance.

2. Answer **either** A **or** B.

    A. Give an account of the problems of osmoregulation in freshwater bony fish and outline their adaptations to overcome these problems.   **(10)**

    OR

    B. Give an account of obtaining food in animals by reference to co-operative hunting, dominance hierarchy, and territorial behaviour.   **(10)**

[END OF QUESTION PAPER]

SPACE FOR ANSWERS

SPACE FOR ANSWERS

Page thirty-two

SPACE FOR ANSWERS

SPACE FOR ANSWERS

SPACE FOR ANSWERS

SPACE FOR ANSWERS

Page thirty-four

SPACE FOR ANSWERS

SPACE FOR ANSWERS

ADDITIONAL GRAPH PAPER FOR QUESTION 10(*b*)

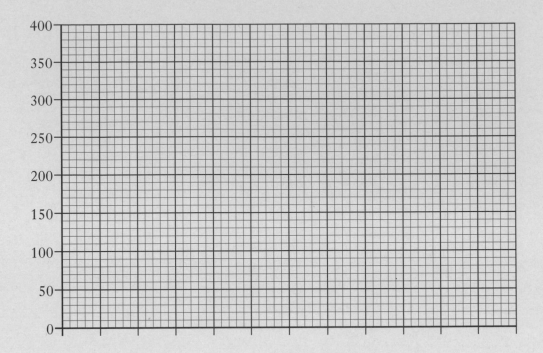

[BLANK PAGE]

FOR OFFICIAL USE

Total for
Sections
B and C

X007/301

NATIONAL
QUALIFICATIONS
2008

TUESDAY, 27 MAY
1.00 PM – 3.30 PM

BIOLOGY
HIGHER

Fill in these boxes and read what is printed below.

Full name of centre

Town

Forename(s)

Surname

Date of birth
  Day   Month   Year   Scottish candidate number   Number of seat

SECTION A—Questions 1–30 (30 marks)

Instructions for completion of Section A are given on page two.

For this section of the examination you must use an **HB pencil**.

SECTIONS B AND C (100 marks)

1   (a)   All questions should be attempted.

     (b)   It should be noted that in **Section C** questions 1 and 2 each contain a choice.

2   The questions may be answered in any order but all answers are to be written in the spaces provided in this answer book, **and must be written clearly and legibly in ink**.

3   Additional space for answers will be found at the end of the book. If further space is required, supplementary sheets may be obtained from the invigilator and should be inserted inside the **front cover** of this book.

4   The numbers of questions must be clearly inserted with any answers written in the additional space.

5   Rough work, if any should be necessary, should be written in this book and then scored through when the fair copy has been written. If further space is required a supplementary sheet for rough work may be obtained from the invigilator.

6   Before leaving the examination room you must give this book to the invigilator. If you do not, you may lose all the marks for this paper.

Read carefully

1   Check that the answer sheet provided is for **Biology Higher (Section A)**.

2   For this section of the examination you must use an **HB pencil**, and where necessary, an eraser.

3   Check that the answer sheet you have been given has **your name**, **date of birth**, **SCN** (Scottish Candidate Number) and **Centre Name** printed on it.

    Do not change any of these details.

4   If any of this information is wrong, tell the Invigilator immediately.

5   If this information is correct, **print** your name and seat number in the boxes provided.

6   The answer to each question is **either** A, B, C or D. Decide what your answer is, then, using your pencil, put a horizontal line in the space provided (see sample question below).

7   There is **only one correct** answer to each question.

8   Any rough working should be done on the question paper or the rough working sheet, **not** on your answer sheet.

9   At the end of the exam, put the **answer sheet for Section A inside the front cover of this answer book**.

Sample Question

The apparatus used to determine the energy stored in a foodstuff is a

A   calorimeter

B   respirometer

C   klinostat

D   gas burette.

The correct answer is **A**—calorimeter. The answer **A** has been clearly marked in **pencil** with a horizontal line (see below).

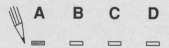

Changing an answer

If you decide to change your answer, carefully erase your first answer and using your pencil fill in the answer you want. The answer below has been changed to **D**.

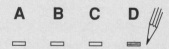

SECTION A

All questions in this section should be attempted.

Answers should be given on the separate answer sheet provided.

1. The following statements relate to respiration and the mitochondrion.

   1 Glycolysis takes place in the mitochondrion.

   2 The mitochondrion has two membranes.

   3 The rate of respiration is affected by temperature.

   Which of the above statements are correct?

   A 1 and 2

   B 1 and 3

   C 2 and 3

   D All of them

2. The anaerobic breakdown of glucose splits from the aerobic pathway of respiration

   A after the formation of pyruvic acid

   B after the formation of acetyl-CoA

   C after the formation of citric acid

   D at the start of glycolysis.

3. Phagocytes contain many lysosomes so that

   A enzymes which destroy bacteria can be stored

   B toxins from bacteria can be neutralised

   C antibodies can be released in response to antigens

   D bacteria can be engulfed into the cytoplasm.

4. After an animal cell is immersed in a hypotonic solution it will

   A burst

   B become turgid

   C shrink

   D become flaccid.

5. Which of the following proteins has a fibrous structure?

   A Pepsin

   B Amylase

   C Insulin

   D Collagen

6. The following cell components are involved in the synthesis and secretion of an enzyme.

   1 Golgi apparatus

   2 Ribosome

   3 Cytoplasm

   4 Endoplasmic reticulum

   Which of the following identifies correctly the route taken by an amino acid molecule as it passes through these cell components?

   A 3 2 1 4

   B 2 4 3 1

   C 3 2 4 1

   D 3 4 2 1

[Turn over

7. The graphs show the effect of various factors on the rate of uptake of chloride ions by discs of carrot tissue from their surrounding solution.

1

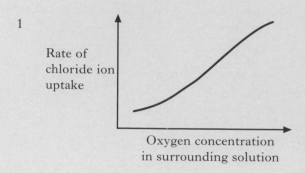

2

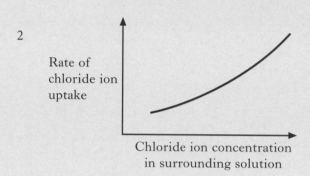

3

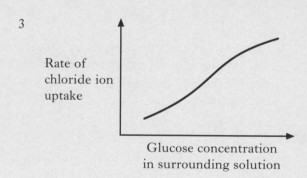

Which graphs support the hypothesis that chloride ion uptake by carrot tissue involves active transport?

A   1 and 2 only

B   1 and 3 only

C   2 and 3 only

D   1, 2 and 3

8. The R_f value of a pigment can be calculated as follows:

$$R_f = \frac{\text{distance travelled by pigment from origin}}{\text{distance travelled by solvent from origin}}$$

The diagram shows a chromatogram in which four chlorophyll pigments have been separated.

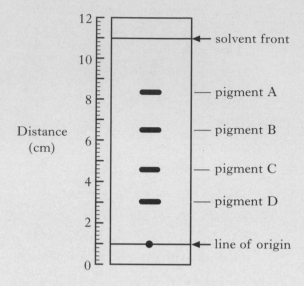

Which pigment has an R_f value of 0·2?

9. The following steps occur during the replication of a virus.

1   alteration of host cell metabolism

2   production of viral protein coats

3   replication of viral nucleic acid

In which sequence do these events occur?

A   1, 3, 2

B   1, 2, 3

C   2, 1, 3

D   3, 1, 2

10. The graphs below show the effect of two injections of an antigen on the formation of an antibody.

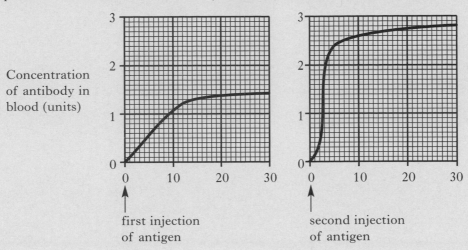

Concentration of antibody in blood (units)

first injection of antigen

second injection of antigen

Time in days

The concentration of antibodies is measured 25 days after each injection. The effect of the second injection is to increase the concentration by

A   1%

B   25%

C   50%

D   100%

11. The diagram shows a stage of meiosis.

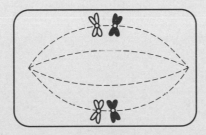

Which of the following diagrams shows the next stage in meiosis?

A

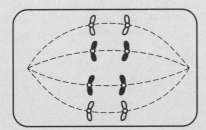

B

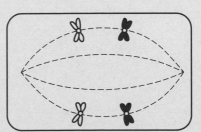

C

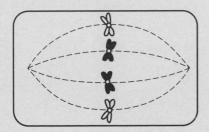

D

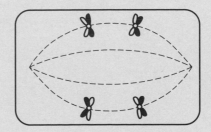

12. Cystic fibrosis is a genetic condition caused by an allele that is not sex-linked. A child is born with cystic fibrosis despite neither parent having the condition. The parents are going to have a second child.

    What is the chance that the second child will have cystic fibrosis?

    A   75%

    B   67%

    C   50%

    D   25%

13. A sex-linked condition in humans is caused by a recessive allele.

    An unaffected man and a carrier woman produce a son.

    What is the chance that he will be unaffected?

    A   1 in 1

    B   1 in 2

    C   1 in 3

    D   1 in 4

14. A new species of organism is considered to have evolved when its population

    A   is isolated from the rest of the population by a geographical barrier

    B   shows increased variation due to mutations

    C   can no longer interbreed successfully with the rest of the population

    D   is subjected to increased selection pressure in its habitat.

15. The melanic variety of the peppered moth became common in industrial areas of Britain following the increase in the production of soot during the industrial revolution.

    The increase in the melanic variety was due to

    A   melanic moths migrating to areas which gave the best camouflage

    B   a change in selection pressure

    C   an increase in the mutation rate

    D   a change in the prey species taken by birds.

16. Which of the following is true of freshwater fish?

    A   The kidneys contain few small glomeruli.

    B   The blood filtration rate is high.

    C   Concentrated urine is produced.

    D   The chloride secretory cells actively excrete excess salts.

17. Which of the following is a behavioural adaptation used by some mammals to survive in hot deserts?

    A   Dry mouth and nasal passages

    B   High levels of anti-diuretic hormone in the blood

    C   Very long kidney tubules

    D   Nocturnal habit

18. Which line in the table below correctly identifies the effect of the state of the guard cells on the opening and closing of stomata?

	State of guard cells	Stomata open/closed
A	flaccid	open
B	plasmolysed	open
C	flaccid	closed
D	turgid	closed

19. In an animal, habituation has taken place when a

    A   harmful stimulus ceases to produce a response

    B   harmful stimulus always produces an identical response

    C   harmless stimulus ceases to produce a response

    D   harmless stimulus always produces an identical response.

20. The table below shows the rate of production of urine by a salmon in both fresh and sea water.

	Rate of urine production (cm³/kg of body mass/hour)
In fresh water	5·0
In sea water	0·5

After transfer from the sea to fresh water, the volume of urine produced by a 2·5 kg salmon over a one hour period would have increased by

A   4·50 cm³

B   5·50 cm³

C   11·25 cm³

D   12·50 cm³.

21. A 30 g serving of a breakfast cereal contains 1·5 mg of iron. Only 25% of this iron is absorbed into the bloodstream.

If a pregnant woman requires a daily uptake of 6 mg of iron, how much cereal would she have to eat each day to meet this requirement?

A   60 g

B   120 g

C   240 g

D   480 g

22. Which of the following graphs represents the growth pattern of a locust?

A

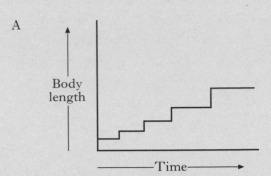

B

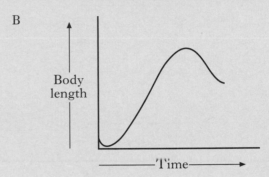

C

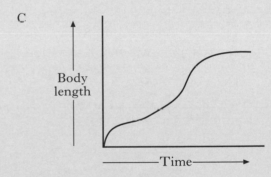

D

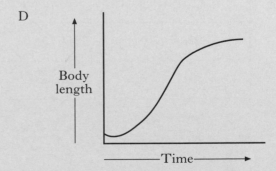

[Turn over

23. Gene expression in cells results in the synthesis of specific proteins. The process of transcription involved in the synthesis of a protein is the

A production of a specific mRNA

B processing of a specific mRNA on the ribosomes

C replication of DNA in the nucleus

D transfer of amino acids to the ribosomes.

24. Hormones P and Q are involved in the control of growth and metabolism.

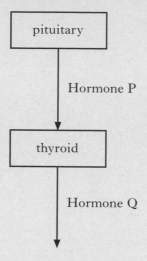

Control of growth and metabolism

Which line in the table below correctly identifies hormones P and Q?

	Hormone P	Hormone Q
A	TSH	growth hormone
B	thyroxine	TSH
C	TSH	thyroxine
D	growth hormone	TSH

25. Temperature control mechanisms in the skin of mammals are stimulated by

A nerve impulses from the pituitary gland

B nerve impulses from the hypothalamus

C hormonal messages from the hypothalamus

D hormonal messages from the pituitary gland.

26. Which of the following is **not** an effect of IAA?

A Increased stem elongation

B Fruit formation

C Inhibition of leaf abscission

D Initiation of germination

27. List P gives reasons why population monitoring may be carried out.

List Q gives three species whose populations are monitored by scientists.

List P
1 Valuable food resource
2 Endangered species
3 Indicator species

List Q
W Stonefly
X Humpback Whale
Y Haddock

Which line in the table below correctly matches reasons from **List P** with species from **List Q**?

	Reasons		
	1	2	3
A	W	X	Y
B	Y	W	X
C	X	Y	W
D	Y	X	W

28. The graph below records the body temperature of a woman during an investigation in which her arm was immersed in water.

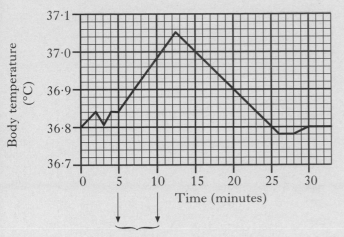

Arm immersed in water during this period

By how much did the temperature of her body vary during the 30 minutes of the investigation?

A 0·25 °C

B 0·27 °C

C 2·5 °C

D 2·7 °C

29. The graph below shows the variation in numbers of a predator and its prey recorded over a ten week period.

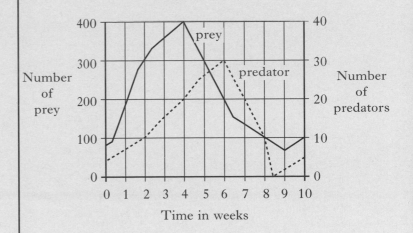

In which week is the prey to predator ratio the largest?

A week 2

B week 4

C week 6

D week 8

30. The graph below shows the length of a human fetus before birth.

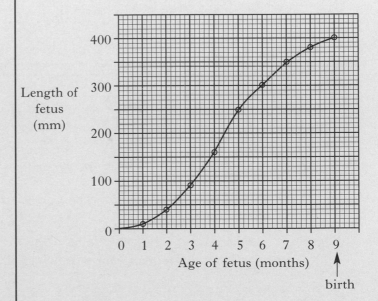

What is the percentage increase in length of the fetus during the 4 months before birth?

A 33·3%

B 37·5%

C 60·0%

D 150%

Candidates are reminded that the answer sheet MUST be returned INSIDE the front cover of this answer book.

Marks

SECTION B

All questions in this section should be attempted.

All answers must be written clearly and legibly in ink.

1. The diagram shows a chloroplast from a palisade mesophyll cell.

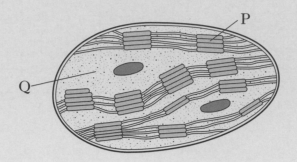

(a) Name regions P and Q.

P _____

Q _____   1

(b) (i) Mark an X on the diagram to show the location of chlorophyll molecules.   1

(ii) As well as chlorophyll, plants have other photosynthetic pigments.

State the benefit to plants of having these other pigments.

_____

_____   1

(c) (i) Name **one** product of the light dependent stage of photosynthesis which is required for the carbon fixation stage (Calvin cycle).

_____   1

(ii) The table shows some substances involved in the carbon fixation stage of photosynthesis.

Complete the table by inserting the number of carbon atoms present in one molecule of each substance.

Substance	*Number of carbon atoms in one molecule*
Glucose	
Carbon dioxide	
Glycerate phosphate (GP)	
Ribulose bisphosphate (RuBP)	

2

Marks

1. **(continued)**

(*d*) The graph below shows the effect of increasing light intensity on the rate of photosynthesis at different carbon dioxide concentrations and temperatures.

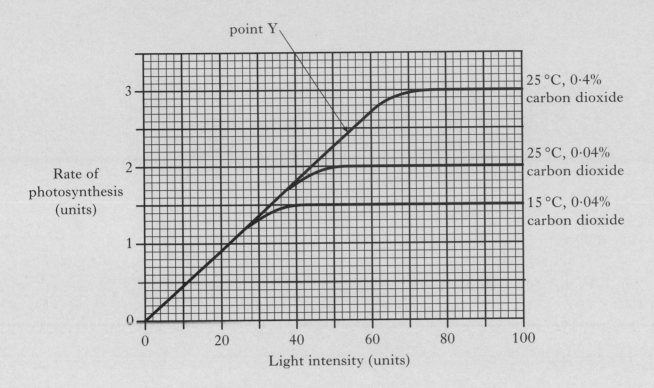

(i) Identify the factor limiting the rate of photosynthesis at point Y on the graph.

_____   1

(ii) From the graph, identify the factor that has the greatest effect in increasing the rate of photosynthesis at a light intensity of 80 units.

Justify your answer.

Factor _____

Justification _____

_____

_____   1

[Turn over

2. The diagram shows a human liver cell and a magnified section of its plasma membrane.

Marks

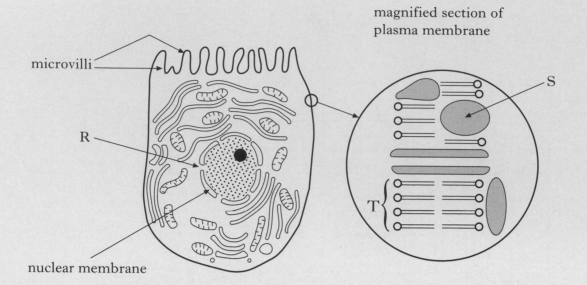

magnified section of
plasma membrane

microvilli

R

nuclear membrane

S

T

(a)   (i)   Identify molecules S and T.

S _____   **1**

T _____   **1**

(ii)   A pore in the nuclear membrane is shown by label R.

Describe the importance of these pores in protein synthesis.

_____

_____   **1**

(iii)   What evidence in the diagram suggests that this cell produces large quantities of ATP?

_____   **1**

DO NOT WRITE IN THIS MARGIN

Marks

2. (continued)

(*b*) Some liver cells take up glucose from the blood by the process of diffusion.

(i) Describe this process.

____

____ **1**

(ii) Suggest a reason for the presence of microvilli in liver cells as shown in the diagram.

____

____

____ **2**

(iii) Glucose taken up by liver cells can be converted into a storage carbohydrate.

Name this carbohydrate.

____________________ **1**

[Turn over

Marks

3. Fat can be used as an alternative respiratory substrate. The diagram shows the breakdown of fat in an athlete's muscle cells during the final stages of a marathon race.

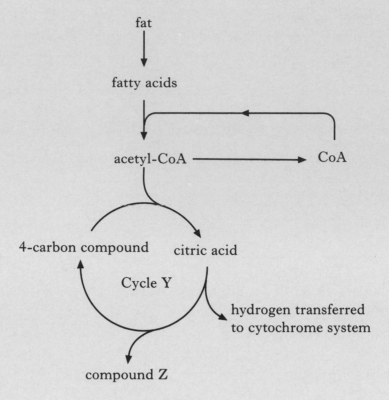

(a) Name a respiratory substrate, other than fat, which can be used by muscle cells.

_____   1

(b) (i) Name cycle Y and compound Z.

Cycle Y _____

Compound Z _____   1

(ii) Name the carrier that accepts and transfers hydrogen to the cytochrome system.

_____   1

(c) Describe the role of oxygen in aerobic respiration.

_____

_____   1

Marks

3. **(continued)**

(*d*) During an 800 metre race, an athlete's muscle cells may respire anaerobically to produce ATP.

(i) State **one** other metabolic product of anaerobic respiration in muscle cells.

_____   1

(ii) Where in a cell does anaerobic respiration occur?

_____   1

(iii) Describe the importance of ATP to cells.

_____

_____   1

[Turn over

3. **(continued)**

Marks

4. (*a*) Decide if each of the following statements about DNA replication is **True** or **False** and tick (✓) the appropriate box.

If the statement is False, write the correct word in the correction box to replace the word underlined in the statement.

Statement	True	False	Correction
During DNA replication <u>hydrogen</u> bonds between bases break.			
During the formation of a new DNA molecule, base pairing is followed by bonding between deoxyribose and <u>bases</u>.			
As a result of DNA replication, the DNA content of a cell is <u>halved</u>.			

2

(*b*) Free DNA nucleotides are needed for DNA replication.

Name **one** other substance that is needed for DNA replication.

_____

1

(*c*) A single strand of a DNA molecule has 6000 nucleotides of which 24% are adenine and 18% are cytosine.

(i) Calculate the combined percentage of thymine and guanine bases on the same DNA strand.

Space for calculation

_____ %   1

(ii) How many guanine bases would be present on the complementary strand of this DNA molecule?

Space for calculation

_____ bases   1

Marks

5. Ponderosa pine trees produce resin following damage to their bark.

In an investigation, three individual pine trees were chosen from areas with different population densities. Each tree was damaged by having a hole bored through its bark.

Measurements of resin production from each hole following this damage are shown in the table.

Population density (Number of trees per hectare)	Volume of resin produced in the first day (cm³)	Duration of resin flow (days)	Total volume of resin produced (cm³)
2	8·3	7·0	29·3
10	0·8	4·8	2·9
50	0·6	4·6	2·8

(*a*) (i) Describe how population density affects the total volume of resin produced.

_____

_____

_____   2

    (ii) Calculate the average resin flow per day at a population density of 2 trees per hectare **after the first day**.

Space for calculation

_____ cm³ per day   1

(*b*) Explain how resin production protects trees.

_____

_____   1

[Turn over

6. An investigation was carried out to compare the rates of water loss from tree species during winter when soil water availability is low.

The table shows information about the tree species involved.

Tree species	Leaf type	Leaves lost in winter
cherry laurel	broad	no
white oak	broad	yes
loblolly pine	needle-like	no

One year old trees of each species were grown outside in identical environmental conditions during winter. The average rate of water loss from each species was measured every tenth day over a 70 day period.

The results are shown in **Graph 1**.

Graph 1

(a) (i) **Use values from Graph 1** to describe the changes in rate of water loss from loblolly pine over the 70 day period.

_____

_____

_____   2

(ii) Calculate the percentage decrease in rate of water loss from cherry laurel between day 0 and day 50.

Space for calculation

_____ %   1

6. (*a*) **(continued)**

(iii) **From Graph 1** express, as the simplest whole number ratio, the rates of water loss from white oak and cherry laurel on day 20.

_____white oak : _____cherry laurel   **1**

(iv) Using the information from the table **and** from Graph 1, suggest the advantage to the white oak of losing its leaves in winter.

Justify your answer.

Advantage _____

Justification _____

_____

_____   **2**

(*b*) In a further investigation, the effect of air temperature on the average rate of water loss from loblolly pine was measured.

The results are shown in **Graph 2**.

Graph 2

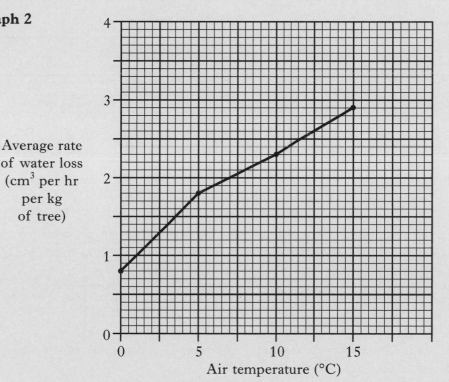

Average rate of water loss (cm^3 per hr per kg of tree)

Air temperature (°C)

(i) Use the information from **Graphs 1 and 2** to suggest the air temperature on day 30 of the investigation.

_____ °C   **1**

(ii) Predict the rate of water loss from loblolly pine at an air temperature of 18 °C.

_____cm^3 per hour per kg of tree.   **1**

(iii) Apart from air temperature and soil water availability, state **one** factor which can affect water loss from trees.

_____   **1**

Marks

7. In horses, coat colour is determined by two genes. The allele for black coat (**B**) is dominant to the allele for chestnut coat (**b**). The allele for grey coat (**G**) is dominant to the allele for non-grey coat (**g**).

Horses with the allele **G** are **always** grey.

A male with the genotype **GgBb** was crossed with a female with the genotype **ggBb**.

(*a*) (i) State the phenotype of each parent.

Male _____

Female _____   **1**

(ii) Complete the grid by adding the genotypes of:

1 the male and female gametes;   **1**

2 the possible offspring.   **1**

	Male gametes		
Female gametes			

(iii) Give the expected phenotype ratio of the offspring from this cross.

_____ Grey : _____ Black : _____ Chestnut   **1**

(*b*) A further gene determines the presence of large white markings in the coat. The allele for the presence of white markings (**T**) is dominant to the allele for their absence (**t**).

A breeder found that a male horse with white markings always produced offspring with white markings when crossed with a female of any phenotype.

Explain this observation in terms of the genotype of this male horse.

_____

_____   **1**

Marks

8. Grey wolves hunt in packs. Their prey includes a variety of large herbivores.

   (*a*) (i) Name the hunting method used by wolves and state **one** advantage of this method.

   Name _____   1

   Advantage _____

   _____   1

   (ii) Following the capture of prey, higher ranking wolves feed first.

   State the term which describes this type of social organisation.

   _____   1

   (iii) Wolf packs occupy territories ranging from 80 to $1500\,km^2$.

   1 Describe **one** advantage to the wolf pack of occupying a territory.

   _____

   _____   1

   2 Suggest **one** factor that could influence the size of a territory occupied by a wolf pack.

   _____   1

   (*b*) The grey wolf was once common in North America but is now an endangered species in many areas.

   Following steps to conserve the species, wolf numbers in one wildlife reserve increased from 31 to 683 individuals during an eight year period.

   (i) Calculate the average yearly increase in wolf numbers during this period.

   Space for calculation

   _____ per year   1

   (ii) Other than wildlife reserves, describe **one** method used to conserve endangered species.

   _____   1

   [Turn over

9. Beech trees have two types of leaf. Sun leaves are exposed to high light intensities for most of the day and shade leaves are usually overshadowed by sun leaves.

The rates of carbon dioxide exchange at different light intensities were measured for sun leaves and shade leaves from one beech tree.

The results are shown on the graph.

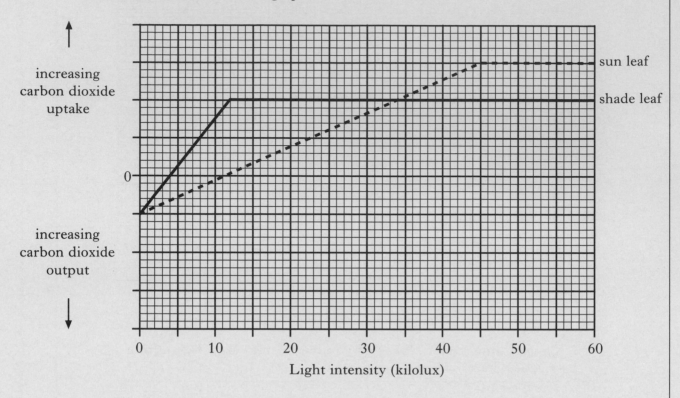

(*a*) State the light intensity at which the shade leaves reach their compensation point.

_____ kilolux   **1**

(*b*) Explain why having shade leaves is an advantage to a beech tree.

_____

_____   **1**

Marks

10. Camels live in deserts where temperatures rise to 50 °C during the day and fall to minus 10 °C at night. The graph shows how the body temperature of a camel varied over a three day period.

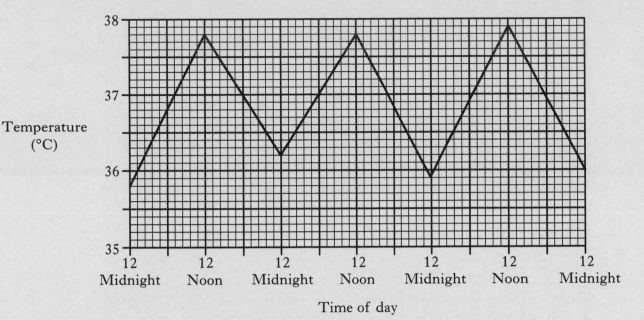

Time of day

(a) From the information given, what evidence is there that camels obtain heat from their own metabolism?

_____

_____   1

(b) What term is used for animals that obtain most of their body heat from their own metabolism?

_____   1

[Turn over

11. In an investigation into the effect of potassium on barley root growth, twelve *Marks* containers were set up as shown.

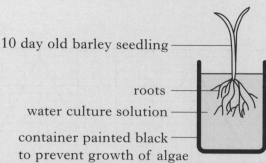

10 day old barley seedling

roots

water culture solution

container painted black
to prevent growth of algae

The water culture solution provided all the elements needed for normal growth.

In six of the containers, the potassium concentration was 2 micromoles per litre. In the other six containers, the potassium concentration was 5 millimoles per litre.

The containers were kept at 20 °C and in constant light intensity.

Every three days, the roots from one container at each potassium concentration were harvested and their dry mass measured.

(a) How many times greater was the potassium concentration in the 5 millimoles per litre solution than in the 2 micromoles per litre solution?

1 millimole per litre = 1000 micromoles per litre

Space for calculation

_____ times   **1**

(b) (i) Identify **one** variable, not already described, that should be kept constant.

_____   **1**

(ii) Suggest **one** advantage of growing the seedlings in water culture solutions rather than soil.

_____   **1**

(iii) Complete the table to give the reasons for each experimental procedure.

Experimental procedure	Reason
paint containers black to prevent growth of algae	
measure dry mass rather than fresh mass of roots	

2

11. (continued)

Marks

(*c*) The results of the investigation are shown in the table.

Time (days)	Dry mass of roots (mg)	
	Potassium concentration 2 micromoles per litre	Potassium concentration 5 millimoles per litre
3	1	1
6	5	6
9	8	10
12	11	14
15	16	22
18	22	44

The results for the seedlings grown in 5 millimoles potassium per litre solution are shown on the graph.

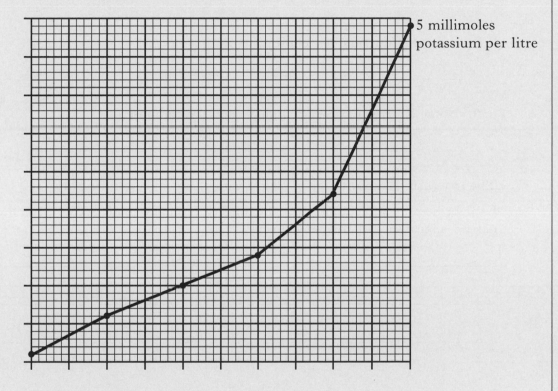

Complete the graph by:

(i) adding the scale and label to each axis;   **1**

(ii) presenting the results for the 2 micromoles potassium per litre solution **and** labelling the line.   **1**

(Additional graph paper, if required, will be found on page 36.)

(*d*) In a further experiment, bubbling oxygen through the water culture solutions was observed to increase the uptake of potassium by the barley roots.

Explain this observation.

_____

_____

_____   **2**

Marks

12. The diagram shows a section through a three year old hawthorn twig with annual rings.

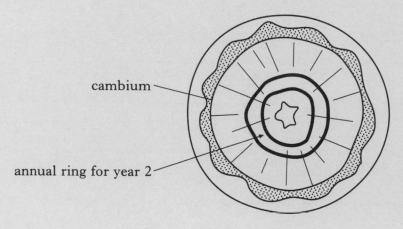

cambium

annual ring for year 2

(a) (i) State the function of cambium.

_____   1

(ii) Name the tissue of which annual rings are composed.

_____   1

(b) The tree suffered an infestation of leaf-eating caterpillars during year 2.

Explain how an infestation of leaf-eating caterpillars could account for the narrow appearance of this annual ring.

_____

_____

_____   2

Marks

13. The flow chart shows part of the homeostatic control of water concentration in human blood.

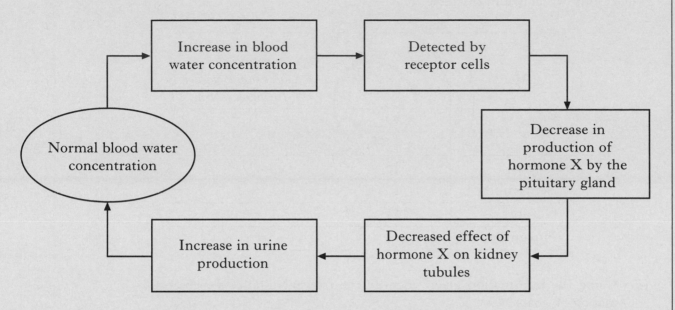

(*a*) (i) Suggest a reason for the increase in blood water concentration.

_____   1

(ii) State the location of the receptor cells.

_____   1

(iii) Name hormone X and state its effect on the kidney tubules.

Name _____   1

Effect _____

_____   1

(*b*) Control of water concentration in human blood involves negative feedback. Explain what is meant by negative feedback.

_____

_____

_____   2

[Turn over

Marks

14. Frequency of mating in a population of wild goats was observed from June to November.

The results are shown in the table.

Month	Average number of hours of light per day	Frequency of mating 0 = no mating + = occasional mating ++ = frequent mating
June	19	0
July	17	0
August	15	0
September	13	+
October	11	++
November	9	++

(a) Using the information given, identify the trigger stimulus which results in mating of goats.

_____ **1**

(b) Young goats are born 5 months after mating.

Explain how the pattern of mating frequency shown increases the survival rate of the offspring.

_____

_____

_____ **2**

(c) What general term is used to describe the effect of light on the timing of breeding in mammals such as goats?

_____ **1**

SECTION C

Both questions in this section should be attempted.

Note that each question contains a choice.

Questions 1 and 2 should be attempted on the blank pages which follow.

Supplementary sheets, if required, may be obtained from the invigilator.

All answers must be written clearly and legibly in ink.

Labelled diagrams may be used where appropriate.

Marks

1. Answer **either** A **or** B.

   A. Write notes on:

   (i) the control of lactose metabolism in *E. coli*;   **6**

   (ii) phenylketonuria in humans.   **4**

   (10)

   OR

   B. Write notes on population change under the following headings:

   (i) the influence of density dependent factors;   **5**

   (ii) succession in plant communities.   **5**

   (10)

In question 2, ONE mark is available for coherence and ONE mark is available for relevance.

2. Answer **either** A **or** B.

   A. Give an account of gene mutations and mutagenic agents.   **(10)**

   OR

   B. Give an account of somatic fusion in plants and genetic engineering in bacteria.   **(10)**

[END OF QUESTION PAPER]

SPACE FOR ANSWERS

SPACE FOR ANSWERS

SPACE FOR ANSWERS

Page thirty-one

SPACE FOR ANSWERS

[Turn over

SPACE FOR ANSWERS

SPACE FOR ANSWERS

SPACE FOR ANSWERS

SPACE FOR ANSWERS

SPACE FOR ANSWERS

SPACE FOR ANSWERS

SPACE FOR ANSWERS

ADDITIONAL GRAPH PAPER FOR QUESTION 11(*c*)

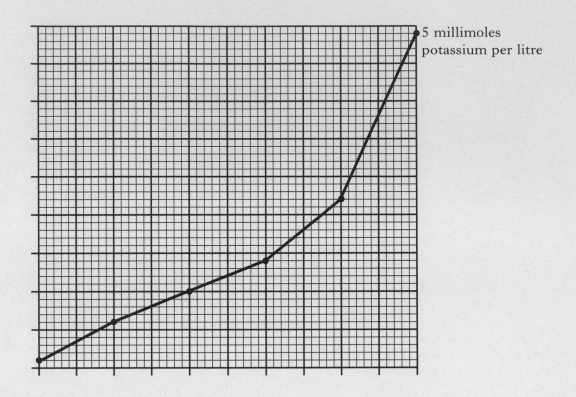

5 millimoles
potassium per litre

HIGHER

2009

[BLANK PAGE]

FOR OFFICIAL USE

Total for
Sections
B and C

X007/301

NATIONAL
QUALIFICATIONS
2009

THURSDAY, 28 MAY
1.00 PM – 3.30 PM

BIOLOGY
HIGHER

Fill in these boxes and read what is printed below.

Full name of centre

Town

Forename(s)

Surname

Date of birth
Day Month Year   Scottish candidate number   Number of seat

SECTION A—Questions 1–30 (30 marks)

Instructions for completion of Section A are given on page two.

For this section of the examination you must use an **HB pencil**.

SECTIONS B AND C (100 marks)

1   (a)   All questions should be attempted.

(b)   It should be noted that in **Section C** questions 1 and 2 each contain a choice.

2   The questions may be answered in any order but all answers are to be written in the spaces provided in this answer book, **and must be written clearly and legibly in ink**.

3   Additional space for answers will be found at the end of the book. If further space is required, supplementary sheets may be obtained from the invigilator and should be inserted inside the **front cover of this book**.

4   The numbers of questions must be clearly inserted with any answers written in the additional space.

5   Rough work, if any should be necessary, should be written in this book and then scored through when the fair copy has been written. If further space is required a supplementary sheet for rough work may be obtained from the invigilator.

6   Before leaving the examination room you must give this book to the invigilator. If you do not, you may lose all the marks for this paper.

Read carefully

1   Check that the answer sheet provided is for **Biology Higher (Section A)**.

2   For this section of the examination you must use an **HB pencil**, and where necessary, an eraser.

3   Check that the answer sheet you have been given has **your name**, **date of birth**, **SCN** (Scottish Candidate Number) and **Centre Name** printed on it.

    Do not change any of these details.

4   If any of this information is wrong, tell the Invigilator immediately.

5   If this information is correct, **print** your name and seat number in the boxes provided.

6   The answer to each question is **either** A, B, C or D. Decide what your answer is, then, using your pencil, put a horizontal line in the space provided (see sample question below).

7   There is **only one correct** answer to each question.

8   Any rough working should be done on the question paper or the rough working sheet, **not** on your answer sheet.

9   At the end of the exam, put the **answer sheet for Section A inside the front cover of this answer book**.

Sample Question

The apparatus used to determine the energy stored in a foodstuff is a

A   calorimeter

B   respirometer

C   klinostat

D   gas burette.

The correct answer is **A**—calorimeter. The answer **A** has been clearly marked in **pencil** with a horizontal line (see below).

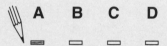

Changing an answer

If you decide to change your answer, carefully erase your first answer and using your pencil fill in the answer you want. The answer below has been changed to **D**.

SECTION A

All questions in this section should be attempted.

Answers should be given on the separate answer sheet provided.

1. Which of the following is **not** surrounded by a membrane?

   A   Nucleus

   B   Ribosome

   C   Chloroplast

   D   Mitochondrion

2. The diagram below shows a plant cell which has been placed in a salt solution.

   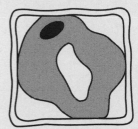

   Which line in the table describes correctly the salt solution and the state of the plant cell?

	Salt solution	*State of cell*
A	hypertonic	plasmolysed
B	hypertonic	turgid
C	hypotonic	flaccid
D	hypotonic	plasmolysed

3. Thin sections of beetroot and rhubarb tissue were immersed in the same sucrose solution for the same time. This resulted in the plasmolysis of 0% of the beetroot cells and 20% of the rhubarb cells.

   Which of the following statements can be deduced from these results?

   A   The sucrose solution was hypertonic to the beetroot cells.

   B   The sucrose solution was hypotonic to the rhubarb cells.

   C   The contents of the beetroot cells were hypotonic to the contents of the rhubarb cells.

   D   The contents of the rhubarb cells were hypotonic to the contents of the beetroot cells.

4. The total sunlight energy landing on an ecosystem is 4 million kilojoules per square metre (kJm^{-2}). Four percent of this is fixed during photosynthesis and five percent of this fixed energy is passed on to the primary consumers. What is the energy intake of the primary consumers?

   A   800 kJm^{-2}

   B   8000 kJm^{-2}

   C   20 000 kJm^{-2}

   D   360 000 kJm^{-2}

5. The diagram below shows a chromatogram of four plant pigments.

   The R_f value of each is calculated by dividing the furthest distance the pigment has moved, by the distance the solvent has moved from the origin.

   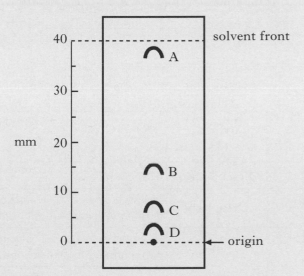

   Which pigment has an R_f value closest to 0·4?

[Turn over

6. The diagram below shows energy transfer within a cell.

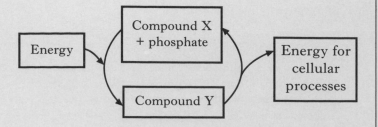

Which line in the table below identifies correctly compounds X and Y?

	X	Y
A	glucose	ATP
B	glucose	ADP
C	ADP	ATP
D	ATP	glucose

7. Which line in the table below shows correctly the sites of stages in aerobic respiration?

	Glycolysis	Krebs Cycle	Cytochrome System
A	Cristae of Mitochondrion	Matrix of Mitochondrion	Cytoplasm
B	Cytoplasm	Cristae of Mitochondrion	Matrix of Mitochondrion
C	Cytoplasm	Matrix of Mitochondrion	Cristae of Mitochondrion
D	Matrix of Mitochondrion	Cytoplasm	Cristae of Mitochondrion

8. Which of the following is **not** composed of amino acids?

A   Glucagon

B   Collagen

C   Amylase

D   Cellulose

9. The table below refers to the mass of DNA in certain human body cells.

Cell type	Mass of DNA in cell $(\times 10^{-12} \text{ g})$
liver	6·6
lung	6·6
P	3·3
Q	0·0

Which of the following is most likely to identify correctly cell types P and Q?

	P	Q
A	kidney cell	sperm cell
B	sperm cell	mature red blood cell
C	mature red blood cell	sperm cell
D	nerve cell	mature red blood cell

10. Which line in the table below identifies correctly cellular defence mechanisms in plants which protect them against micro-organisms and herbivores?

	Defence against micro-organisms	Defence against herbivores
A	antibodies	resins
B	tannins	cyanide
C	spines	cyanide
D	resins	antibodies

11. In poultry, males have two X chromosomes and females have one X chromosome and one Y chromosome.

The gene for feather-barring is sex-linked.

The allele for barred feathers is dominant to the allele for non-barred feathers.

A non-barred male is crossed with a barred female.

What ratio of offspring would be expected?

A 1 barred male : 1 barred female

B 1 non-barred male : 1 non-barred female

C 1 barred male : 1 non-barred female

D 1 non-barred male : 1 barred female

12. The table below shows some genotypes and phenotypes associated with a form of anaemia.

Genotype	Phenotype
AA	Unaffected
AS	Sickle cell trait
SS	Acute sickle cell anaemia

A person with sickle cell trait and an unaffected person have a child together.

What are the chances of the child having acute sickle cell anaemia?

A none

B 1 in 4

C 1 in 2

D 1 in 1

13. Which of the following statements refers to a gene mutation?

A A change in the chromosome number caused by non-disjunction.

B A change in the number of genes on a chromosome caused by duplication.

C A change in the structure of a chromosome caused by translocation.

D A change in the base sequence of DNA caused by substitution.

14. Polyploidy in plants may result from

A total spindle failure during meiosis

B hybridisation between varieties of the same species

C homologous chromosomes binding at chiasmata

D the failure of linked genes to separate.

15. Which of the following is an example of artificial selection?

A Industrial melanism in moths

B DDT resistance in mosquitoes

C Increased milk yield in dairy cattle

D Decreasing effect of antibiotics on bacteria

16. The diagram below shows stages involved in the genetic engineering of bacteria to produce human insulin.

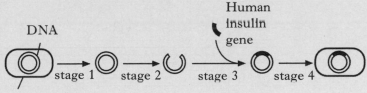

Which line in the table below shows the stages of this process in which endonuclease and ligase are involved?

	Stage involving endonuclease	Stage involving ligase
A	2	4
B	2	3
C	3	2
D	4	3

[Turn over

17. The statements below describe methods of maintaining a water balance in fish.

   1   Salts actively absorbed by chloride secretory cells

   2   Salts actively secreted by chloride secretory cells

   3   Low rate of kidney filtration

   4   High rate of kidney filtration

   Which of these are used by **freshwater** bony fish?

   A   1 and 3 only

   B   2 and 4 only

   C   1 and 4 only

   D   2 and 3 only

18. The graph below shows the net energy gain or loss from hunting and eating prey of different masses.

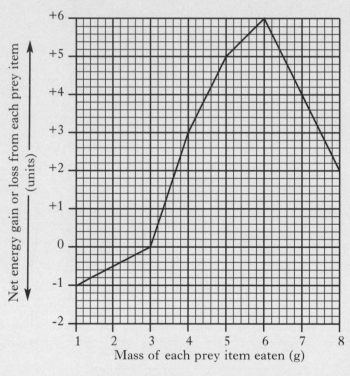

   It can be concluded from the graph that

   A   prey between 1 g and 3 g are rarer than prey between 3 g and 6 g

   B   hunting and eating prey above 6 g involves a net energy loss

   C   prey of 8 g contain less energy than prey of mass 6 g

   D   hunting and eating prey below 3 g involves a net energy loss.

19. Which of the following statements about habituation is correct?

   A   It is a temporary change in behaviour.

   B   It occurs only in young animals.

   C   It is a social mechanism for defence.

   D   It is a permanent change in behaviour.

20. Some animal species live in social groups for defence.

   Which of the following statements describes a change which could result from an increase in the size of such a social group?

   A   Individuals are able to spend less time feeding.

   B   There are fewer times when more than one animal is looking for predators.

   C   Each animal can spend more time looking for predators than foraging.

   D   Individuals are able to spend less time looking for predators.

21. Phenylketonuria is a condition that results from

   A   differential gene expression

   B   chromosome non-disjunction

   C   a vitamin deficiency

   D   an inherited gene mutation.

22. The plant growth substance indole acetic acid (IAA) is of benefit to humans because it can function

   A   as a herbicide and to break dormancy

   B   as a herbicide and as a rooting powder

   C   in the germination of barley and to break dormancy

   D   as a rooting powder and in the germination of barley.

23. The graph below shows changes in the α-amylase concentration and starch content of a barley grain during early growth and development.

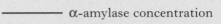

- - - - - - starch content

——————— α-amylase concentration

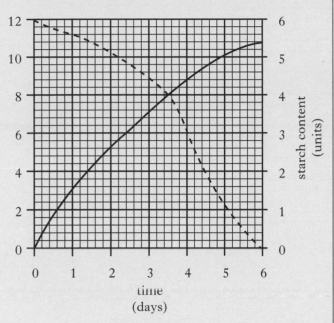

What is the α-amylase concentration when the starch content has decreased by 50%?

A 4·4 units

B 6·0 units

C 8·2 units

D 8·8 units

24. A species of plant was exposed to various periods of light and dark, after which its flowering response was observed.

The results are shown below.

Light period (hours)	Dark period (hours)	Flowering response
4	12	flowering
4	10	flowering
6	18	maximum flowering
14	10	flowering
18	9	no flowering
18	6	no flowering
18	10	flowering

What appears to be the critical factor which stimulates flowering?

A A minimum dark period of 10 hours

B A light/dark cycle of at least 24 hours

C A light period of less than 18 hours

D A dark period which exceeds the light period

25. When there is a decrease in the water concentration of the blood, which of the following series of events shows the negative feedback response of the body?

	Concentration of ADH	Permeability of kidney tubules	Volume of urine
A	increases	increases	increases
B	decreases	decreases	increases
C	increases	increases	decreases
D	decreases	increases	decreases

[Turn over

26. High levels of blood glucose can cause clouding of the lens in the human eye. Concentrations above 5·5 mM are believed to put the individual at a high risk of lens damage.

In an investigation, people of different ages each drank a glucose solution. The concentration of glucose in their blood was monitored over a number of hours. The results are shown in the graph below.

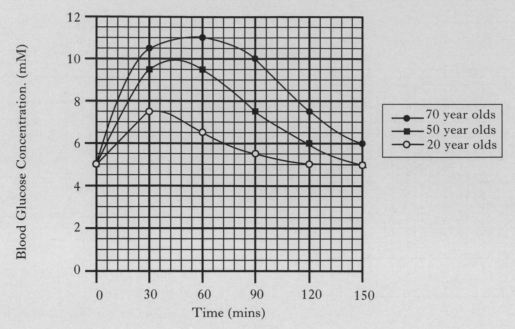

For how long during the investigation did 20 year olds remain above the high risk blood glucose concentration?

A 84 mins

B 90 mins

C 120 mins

D 148 mins

27. Which of the following shows correct responses to changes in sugar concentration in the blood?

	Sugar concentration in blood	Glucagon secretion	Insulin secretion	Glycogen stored in liver
A	increases	decreases	increases	increases
B	increases	decreases	increases	decreases
C	decreases	increases	decreases	increases
D	decreases	decreases	increases	decreases

28. A person produces 0·75 litres of urine in 24 hours. This urine contains 18 g of urea.

What is the concentration of urea in this urine?

A $1·0 \text{ g}/100 \text{ cm}^3$

B   2·4 g/litre

C $2·4 \text{ g}/100 \text{ cm}^3$

D $3·6 \text{ g}/100 \text{ cm}^3$

29. The list below describes changes involved in temperature regulation.

List

1   Increased vasodilation

2   Decreased vasodilation

3   Hair erector muscles contract

4   Hair erector muscles relax

Which of these are responses to cooling in mammals?

A   1 and 3 only

B   1 and 4 only

C   2 and 3 only

D   2 and 4 only

30. Which line in the table below shows correctly the main source of body heat and the method of controlling body temperature in an ectotherm?

	Main source of body heat	*Method of controlling body temperature*
A	Respiration	Physiological
B	Respiration	Behavioural
C	Absorbed from environment	Physiological
D	Absorbed from environment	Behavioural

Candidates are reminded that the answer sheet MUST be returned INSIDE the front cover of this answer book.

[Turn over

Marks

SECTION B

All questions in this section should be attempted.

All answers must be written clearly and legibly in ink.

1. (*a*) The diagram below shows light striking a green leaf.

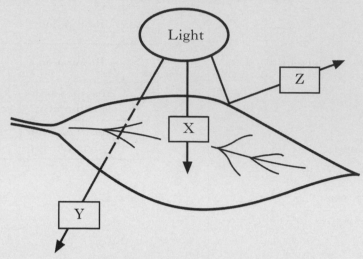

Arrow X shows light being absorbed.

State the terms used to describe what is happening to light at Y and Z.

Y _____

Z _____   1

(*b*) The diagram below represents the absorption of different colours of light by a photosynthetic pigment.

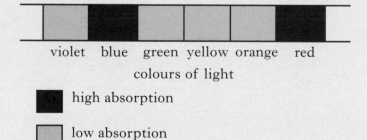

violet   blue   green   yellow   orange   red
colours of light

■ high absorption

▨ low absorption

(i) Name this photosynthetic pigment.

_____   1

(ii) State the role of accessory pigments in photosynthesis.

_____

_____   1

DO NOT
WRITE IN
THIS
MARGIN

Marks

1. **(continued)**

   (c) The diagram below shows an outline of the carbon fixation stage of photosynthesis.

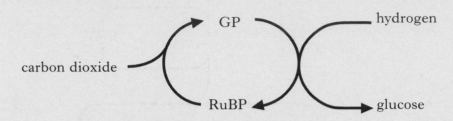

      (i) State the exact location of this stage in a plant cell.

      _____   1

      (ii) Describe the role of hydrogen in the carbon fixation stage.

      _____

      _____   1

   (d) The graph below shows the effect of increasing the concentration of carbon dioxide on the rate of photosynthesis by a plant at different temperatures.

      Light intensity was kept constant.

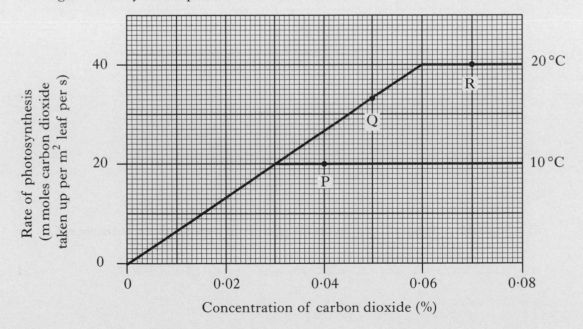

      Using the information in the graph, identify the factor which is limiting the rate of photosynthesis at each of the points P, Q and R.

      P _____

      Q _____

      R _____   2

Marks

2. (*a*) The diagram below shows some of the steps in respiration.

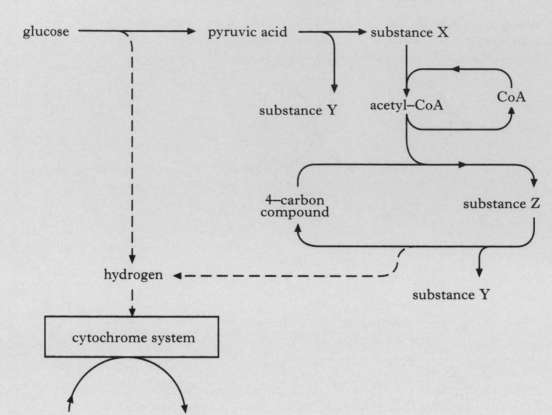

(i) Complete the table below by naming substances W, X and Y.

Substance	Name
W	
X	
Y	

2

(ii) State the number of carbon atoms present in a molecule of substance Z.

_____

1

Marks

2. **(continued)**

(b) Yeast cells were grown in both aerobic and anaerobic conditions and the volume of carbon dioxide produced was measured.

The results are shown in the graph below.

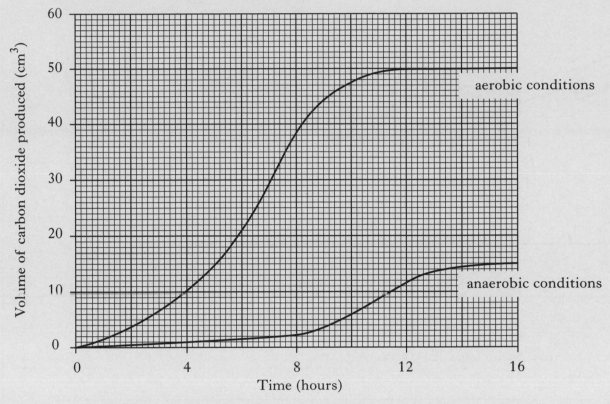

(i) At which time is there the greatest difference between the volumes of carbon dioxide produced in aerobic and anaerobic conditions?

Tick (✓) the correct box.

☐ ☐ ☐ ☐ ☐

8 hours   10 hours   12 hours   14 hours   16 hours

1

(ii) Calculate the average rate of carbon dioxide production per hour over the first 6 hours in aerobic conditions.

Space for calculation

_____ cm³ per hour   1

[Turn over

Marks

3. (*a*) The diagram below shows one stage in the synthesis of a protein at a ribosome.

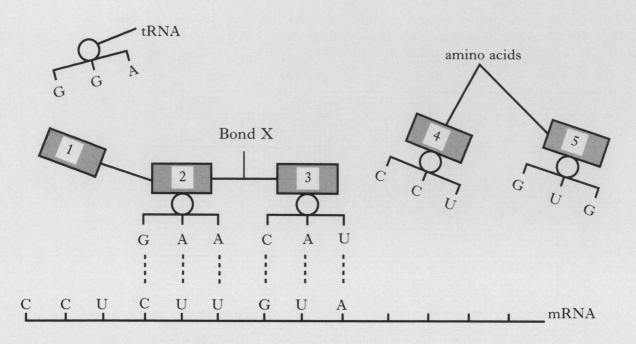

(i) Name this stage in protein synthesis.

_____   1

(ii) Name bond X.

_____   1

(iii) The table below shows five codons and their corresponding amino acids.

Codon	*Amino acid*
CUU	leucine
GGA	glycine
CAA	glutamic acid
GUA	valine
CCU	proline

Use information from the table to identify amino acids 1 and 4.

1 _____

4 _____   1

Marks

3.   (continued)

(*b*)   The diagram below shows a cell from the pancreas.

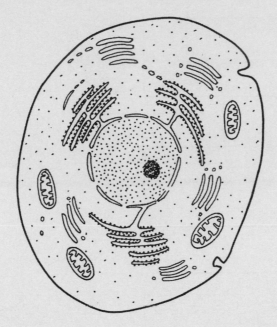

What feature of this cell shows that it is involved in the secretion of protein?

_____ 1

[Turn over

4. *(a)* The diagram below shows some stages during the invasion of a cell by a virus.

Marks

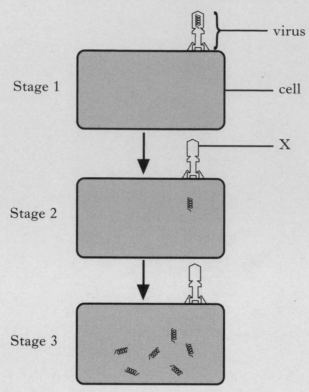

(i) Name the substance of which part X is composed.

_____   **1**

(ii) Describe what happens in the cell between stage 2 and stage 3 to allow the viral nucleic acid to replicate.

_____

_____   **1**

(iii) Describe **two** events which occur between stage 3 and the bursting of the cell to release new viruses.

1 _____

_____

2 _____

_____   **2**

Marks

4. (continued)

(b) The presence of viruses in the human body triggers antibody production by lymphocytes.

(i) What name is given to any substance that triggers this response?

_____ 1

(ii) A person was injected with a vaccine on day 1 and again on day 36 of a 70 day study. The table below shows the concentration of antibodies to this vaccine in this person's blood at the end of each 7 day period during the study.

1st injection   2nd injection

Day	7	14	21	28	35	42	49	56	63	70
Concentration of antibody (mg/100 ml blood)	3	15	28	32	10	80	102	112	120	118

1 How many times greater was the maximum antibody concentration following the second injection compared with the maximum concentration following the first?

Space for calculation

_____ times   1

2 The second injection caused a higher concentration of antibody to be produced than the first injection.

Identify **two** other differences in the response to the second injection.

1 _____

2 _____ 1

[Turn over

5.   Mexican spotted owls are territorial and prey on several species of small mammal.   *Marks*
Three pairs of owls were studied over a two year period.

The table below shows the number of each prey species eaten by each pair of owls.

Prey species	Number of each prey species eaten by each pair of owls		
	Owl pair A	Owl pair B	Owl pair C
deer mouse	484	528	515
woodrat	29	144	141
brush mouse	15	114	118
rock squirrel	22	24	23

The graph below shows the average number and average total biomass of deer mice and woodrats living in the study area in different seasons.

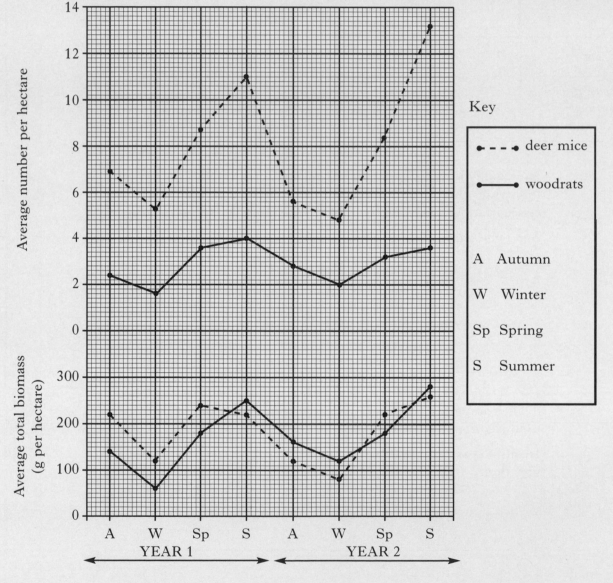

Key

- - - deer mice

——— woodrats

A   Autumn

W   Winter

Sp   Spring

S   Summer

(*a*)   (i)   What percentage of the total number of prey eaten by owl pair A were deer mice?

Space for calculation

_____ %   1

[X007/301]   *Page eighteen*

5. **(a)** **(continued)**

Marks

(ii) Express as the simplest whole number ratio the numbers of deer mice, brush mice and rock squirrels eaten by owl pair B.

Space for calculation

deer mice   :   brush mice   :   rock squirrels

_____ : _____ : _____   1

(b) Use evidence from the table to identify the owl pair that foraged in a different habitat to the other two pairs of owls.

Justify your answer.

Pair _____

Justification _____

_____

_____   1

(c) (i) **Use values from the graph** to describe the change in average number of woodrats per hectare from spring of Year 1 until spring of Year 2.

_____

_____   2

(ii) Calculate the percentage decrease in the average total biomass of woodrats between summer of Year 1 and winter of Year 2.

Space for calculation

_____ %   1

(d) Calculate the average biomass of one deer mouse in summer of Year 1.

Space for calculation

_____ g   1

(e) The size of the territory of a pair of Mexican spotted owls is different in winter and summer. Give an explanation of this observation which can be supported by evidence from the graph.

_____

_____

_____   1

Marks

6. The diagram below shows a pair of homologous chromosomes in a mouse cell during meiosis. The positions of three genes R, S and T are also shown.

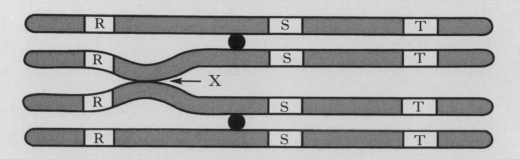

(a) Name an organ in mice where meiosis occurs.

_____ 1

(b) (i) Name point X where crossing over may occur.

_____ 1

(ii) Crossing over leads to recombination of genes.

Between which two genes in the diagram would the greatest frequency of recombination take place?

_____ and _____ 1

(iii) Crossing over is a source of genetic variation.

Name **one** other feature of meiosis which leads to genetic variation.

_____ 1

(c) A mouse egg contains 20 chromosomes.

State the number of chromosomes present in a mouse gamete mother cell.

_____ chromosomes 1

Marks

7. In Labrador dogs, the alleles **B** and **b** and alleles **E** and **e** are involved in the determination of coat colour.

- Labradors with alleles **B** and **E** are always black.

- Labradors with alleles **bb** and **E** are always chocolate coloured.

- Labradors with alleles **ee** are always yellow.

(*a*) A male black Labrador of genotype **BbEe** was crossed with a yellow female of the genotype **bbee**.

(i) Complete the table below to show the genotypes of the gametes of the male.

Genotypes of male gametes			

1

(ii) Give the expected phenotype ratio of the offspring from this cross.

Space for calculation

_____ Black : _____ Chocolate :_____ Yellow

1

(*b*) Give the genotype of a male Labrador which could be crossed with a female of genotype **bbee** to ensure that **all** the offspring produced would be chocolate coloured.

Space for calculation

Genotype _____

1

[Turn over

Marks

8. Cuticles are waxy layers on the surfaces of the leaves of many plant species.

The table below shows the average cuticle thickness of the leaves of five plant species and the rates of water loss through their cuticles at 20 °C with no air movement.

Species	Average cuticle thickness (micrometres)	Rate of water loss through cuticle (cubic micrometres per cm^2 per hour)
A	1·4	36·7
B	2·8	25·8
C	4·2	18·1
D	5·6	8·5
E	7·0	8·4

(a) (i) Describe the relationship between average cuticle thickness and rate of water loss through the cuticles in these plant species.

_____

_____

_____   2

(ii) Leaves lose most water through their open stomata.

Give the term used to describe the condition of the guard cells when stomata are open.

_____   1

(iii) State **two** changes to environmental conditions which could lead to an increase in water loss from leaves.

1 _____

2 _____   1

DO NOT WRITE IN THIS MARGIN

Marks

8. (continued)

(*b*) Plants which grow in extremely dry conditions have leaf adaptations which reduce water loss.

(i) Complete the table below to explain how each leaf adaptation reduces water loss.

Leaf adaptation	*Explanation of how the adaptation reduces water loss from leaves*
Presence of hairs on leaf surface	
Leaves small and few in number	

2

(ii) What term describes plants that have adaptations to reduce water loss?

______________________________ 1

[Turn over

Marks

9.   (*a*)   In an investigation into the effects of grazing, the total biomass of grass species and the diversity of all plant species in a field was monitored over a period of four years. The field was not grazed in years 1 and 2. Sheep grazed the field in years 3 and 4.

The results are shown on the graph below.

– – – – diversity   —— biomass

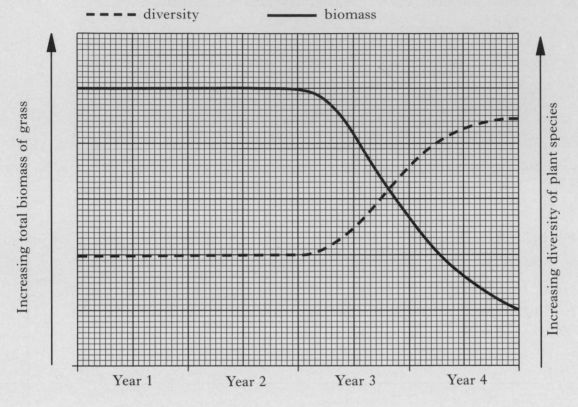

(i)   Explain the effects of grazing by sheep on the total biomass of grass species and the diversity of plant species during year 3.

Total biomass of grass species.

_____   1

Diversity of all plant species

_____

_____   1

(ii)   The number of sheep grazing the field was increased after year 4. Suggest how this would affect the diversity of plant species in the field.

Justify your answer.

Effect on diversity

_____

Justification

_____

_____   1

Marks

9. (continued)

(b) The table below contains statements describing plant adaptations.

Complete the table by ticking (✓) the boxes to show the adaptations that help the plants tolerate the effects of grazing.

Plant adaptation	Tick (✓)
Dandelions have deep roots	
Wild roses have thorns	
Couch grass has underground stems	
Nettles have stings	
Tobacco plants produce nicotine	

2

[Turn over

Marks

10. Catechol oxidase is an enzyme found in apple tissue. It is involved in the reaction which produces the brown pigment that forms in cut or damaged apples.

catechol
(colourless substance   catechol oxidase
in apple tissue) $\xrightarrow{\hspace{3cm}}$ brown pigments

The effect of the concentration of lead ethanoate on this reaction was investigated.

10 g of apple tissue was cut up, added to $10\,cm^3$ of distilled water and then liquidised and filtered. This produced an extract containing both catechol and catechol oxidase.

Test tubes were set up as described in **Table 1** and kept at 20 °C in a water bath.

Table 1

Tube	Contents of tubes
A	sample of extract + $1\,cm^3$ distilled water
B	sample of extract + $1\,cm^3$ 0·01% lead ethanoate solution
C	sample of extract + $1\,cm^3$ 0·1% lead ethanoate solution

Every 10 minutes, the tubes were placed in a colorimeter which measured how much brown pigment was present.

The more brown pigment present the higher the colorimeter reading.

The results are shown in **Table 2**.

Table 2

Time (minutes)	Colorimeter reading (units)		
	Tube A	Tube B	Tube C
	sample of extract + distilled water	sample of extract + 0·01% lead ethanoate	sample of extract + 0·1% lead ethanoate
0	1·6	1·8	1·6
10	7·0	5·0	2·0
20	9·0	6·0	2·2
30	9·6	6·4	2·4
40	10·0	7·0	2·4
50	10·0	7·6	2·4
60	10·0	7·6	2·4

(a) (i) Identify **two** variables not already mentioned that would have to be kept constant.

1 _____   1

2 _____   1

Marks

10. **(a) (continued)**

(ii) Describe how tube A acts as a control in this investigation.

_____

_____   1

(b) Explain why the initial colorimeter readings were not 0·0 units.

_____

_____   1

(c) The results for the extract with 0·1% lead ethanoate are shown in the graph below.

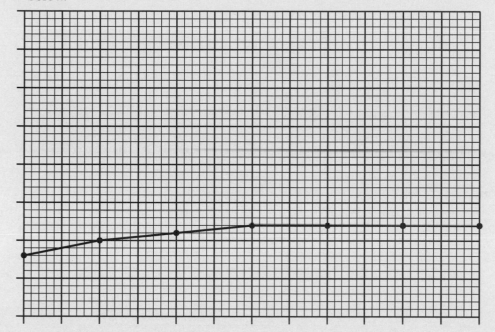

extract + 0·1% lead ethanoate

Use information from **Table 2** to complete the graph by:

(i) adding the scale and label to each axis;   1

(ii) presenting the results for the extract + 0·01% lead ethanoate solution **and** labelling the line.   1

(Additional graph paper, if required, will be found on page 40.)

(d) State the effect of the concentration of lead ethanoate solution on the activity of catechol oxidase.

_____

_____   1

(e) The experiment was repeated with the 0·1% lead ethanoate solution at 60 °C. Predict the colorimeter reading at 10 minutes and justify your answer.

Prediction _____ units

Justification _____

_____   1

Marks

11. Many species of cichlid fish are found in Lake Malawi in Africa.

The diagram below shows the heads of three different cichlid fish and gives information on their feeding methods.

These species have evolved from a single species.

Sucks in microscopic
organisms from the water

Scrapes algae from the
surfaces of rocks

Crushes snail shells and
extracts flesh

(a) Describe how the information given about these fish illustrates adaptive radiation.

_____

_____

_____   **2**

(b) What evidence would confirm that the cichlids are different species?

_____

_____   **1**

(c) The evolution of these cichlid fish has involved geographical isolation.

 (i) Name another type of isolating mechanism.

_____   **1**

 (ii) State the importance of isolating mechanisms in the evolution of new species.

_____

_____   **1**

Marks

12. Diagram A shows a section through a woody stem. Diagram B shows a magnified view of the area indicated on the section.

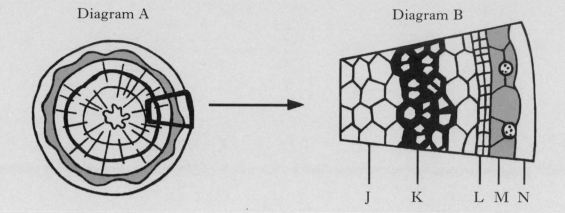

Diagram A

Diagram B

J   K   L M N

(a) Which letter on diagram B shows the position of a lateral meristem?

Letter _____

1

(b) Name the tissue of which annual rings are composed.

_____

1

(c) In which season was the woody stem cut?
Explain your choice.

Season _____

1

Explanation _____

_____

_____

1

[Turn over

Marks

13. The diagram below shows information relating to the Jacob–Monod hypothesis of the control of gene action in the bacterium *Escherichia coli*.

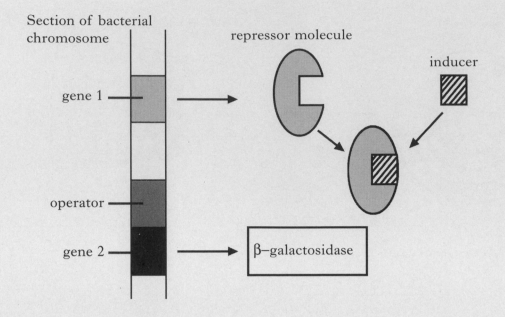

(*a*) Name gene 1.

_____   1

(*b*) Name the substance which acts as the inducer.

_____   1

(*c*) (i) Describe the sequence of events that occurs in the **absence** of the inducer.

_____

_____

_____

_____   2

(ii) Explain why it is important for *E. coli* to control gene action.

_____

_____   1

DO NOT WRITE IN THIS MARGIN

Marks

14. Environmental factors influence growth and development in animals.

(*a*) (i) Explain the importance of iron in the growth and development of humans.

____

______________________________ **1**

(ii) Describe the effects of nicotine on growth and development of a human fetus.

____

______________________________ **1**

(*b*) Breeding behaviour in red deer starts in autumn.

(i) Describe the environmental influence that triggers the start of breeding at this time of year.

____

______________________________ **1**

(ii) Suggest an advantage to red deer of starting to breed at this time of year.

____

______________________________ **1**

[Turn over

Marks

15. The diagram below shows the plant communities that have developed around a fresh water loch.

Fresh water loch

soil

| Pioneer community | Community X | Community Y | Climax community |

Increasing age of communities ⟶

(a) What term describes the process of gradual formation of a climax community?

_____ 1

(b) Suggest a modification that community X may make to its habitat which allows colonisation by community Y.

_____

_____ 1

(c) **Underline** one alternative in each pair to make the sentences correct.

The complexity of the food web in the climax community will be

$\left\{ \begin{array}{l} \text{greater than} \\ \text{less than} \end{array} \right\}$ that in the pioneer community.

Greater species diversity will exist in $\left\{ \begin{array}{l} \text{community X} \\ \text{community Y} \end{array} \right\}$. 1

Marks

SECTION C

Both questions in this section should be attempted.

Note that each question contains a choice.

Questions 1 and 2 should be attempted on the blank pages which follow.

Supplementary sheets, if required, may be obtained from the invigilator.

All answers must be written clearly and legibly in ink.

Labelled diagrams may be used where appropriate.

1. Answer **either** A **or** B.

   A. Write notes on:

   (i)   structure of the plasma membrane;   **4**

   (ii)   function of the plasma membrane in active transport;   **3**

   (iii) structure and function of the cell wall.   **3**

   (10)

   OR

   B. Write notes on :

   (i)   the structure of DNA;   **6**

   (ii)   DNA replication and its importance.   **4**

   (10)

In question 2, ONE mark is available for coherence and ONE mark is available for relevance.

2. Answer **either** A **or** B.

   A. Give an account of the importance of nitrogen, phosphorus and magnesium in plant growth and describe the symptoms of their deficiency.   **(10)**

   OR

   B. Give an account of how animal populations are regulated by density-dependent and by density-independent factors.   **(10)**

[END OF QUESTION PAPER]

[Turn over

SPACE FOR ANSWERS

DO NOT WRITE IN THIS MARGIN

SPACE FOR ANSWERS

SPACE FOR ANSWERS

SPACE FOR ANSWERS

SPACE FOR ANSWERS

DO NOT
WRITE IN
THIS
MARGIN

SPACE FOR ANSWERS

SPACE FOR ANSWERS

ADDITIONAL GRAPH PAPER FOR QUESTION 10(*c*)

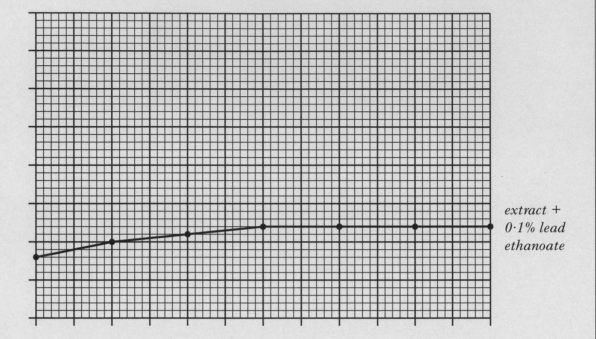

*extract +
0·1% lead
ethanoate*